Understanding the Digital World

Also by Brian W. Kernighan

The Elements of Programming Style (with P. J. Plauger)

Software Tools (with P. J. Plauger)

Software Tools in Pascal (with P. J. Plauger)

The C Programming Language (with Dennis Ritchie)

The AWK Programming Language (with Al Aho and Peter Weinberger)

The Unix Programming Environment (with Rob Pike)

AMPL: A Modeling Language for Mathematical Programming (with Robert Fourer and David Gay)

The Practice of Programming (with Rob Pike)

D is for Digital

Hello, World: Opinion columns from the Daily Princetonian

The Go Programming Language (with Alan Donovan)

Millions, Billions, Zillions: Defending Yourself in a World of Too Many Numbers

Unix: A History and a Memoir

Understanding the Digital World

What You Need to Know about Computers, the Internet, Privacy, and Security

Second Edition

Brian W. Kernighan

Princeton University Press
Princeton and Oxford

Published by Princeton University Press
41 William Street, Princeton, New Jersey 08540
6 Oxford Street, Woodstock, Oxfordshire OX20 1TR

press.princeton.edu

ISBN 978-0-691-21909-7
ISBN (pbk.) 978-0-691-21910-3
ISBN (e-book) 978-0-691-21896-0

British Library Cataloging-in-Publication Data is available.

This book has been composed in Times, Courier and Helvetica using groff, ghostscript, and other open source Unix tools.

The publisher would like to acknowledge the author of this volume for providing the print-ready files from which this book was printed.

Printed on acid-free paper. ∞

Printed in the United States of America

10 9 8 7 6 5 4 3 2

For Meg

Contents

Preface

Almost every fall since 1999, I have taught a Princeton course called "Computers in Our World." The course title is embarrassingly vague, but I had to invent it in less than five minutes one day and then it became too hard to change. Teaching the course itself, however, has proven to be the most fun thing that I do, in a job that is almost entirely enjoyable.

The course is based on the observation that computers and computing are all around us. Some computing is highly visible: every student has a laptop computer that is far more powerful than the single IBM 7094 computer that cost several million dollars, occupied a large air-conditioned room, and served the whole Princeton campus when I arrived there as a graduate student in 1964. Every student has a cell phone too, also with much more computing power than that ancient computer. Every student has high-speed Internet access, as does a significant fraction of the world's population. Everyone searches and shops online, and uses email, texting and social networks to keep in touch with friends and family.

But this is only part of a computing iceberg, much of which lies hidden below the surface. We don't see and usually don't think about the computers that lurk within appliances, cars, airplanes and the pervasive electronic gadgets that we take for granted—smart TVs and thermostats and doorbells, voice recognizers, fitness trackers, earbuds, toys, and games. Nor do we think much about the degree to which infrastructure depends on computing: the telephone network, cable television, air traffic control, the power grid, and banking and financial services.

Most people will not be directly involved in creating such systems, but everyone is strongly affected by them, and some will have to make important decisions about them. An educated person ought to know at least the rudiments of computing: what computers can do and how they do it; what they can't do at all and what's merely too hard right now; how they talk to each other and what happens when they do; and the many ways that computing and communications influence the world around us.

The pervasive nature of computing affects us in unexpected ways. Although we are from time to time reminded of the growth of surveillance systems, incursions into our privacy, and the perils of identity theft, we perhaps do not realize the extent to which they are enabled by computing and communications.

In June 2013, Edward Snowden, a contractor at the United States National Security Agency (NSA), provided journalists with fifty thousand documents which revealed that the NSA had been routinely monitoring and collecting the electronic communications—phone calls, texts, email, and Internet use—of pretty much everyone in the world, but notably of American citizens living in the US who were no threat whatsoever to the security of their country. The Snowden documents also showed that other countries were spying on their citizens too. Perhaps most surprising is that after initial outrage, it's back to business as usual, with more and more government monitoring and spying, and resigned or oblivious acceptance by citizens.

Corporations also track and monitor what we do online and in the real world. The business models of many companies are based on extensive data collection and the ability to predict and influence our behavior. The availability of voluminous data has enabled great progress in speech understanding, computer vision and language translation, but it has come at a cost to our privacy, and has made it hard for anyone to be anonymous.

Hackers of all stripes have become sophisticated in their attacks on data repositories. Electronic break-ins at businesses and government agencies are an almost daily occurrence; information about customers and employees is stolen in large quantities, often to be used for fraud and identity theft. Attacks on individuals are common as well. It used to be that one could be safe from online scams by ignoring mail from putative Nigerian princes or their relatives, but targeted attacks are now far more subtle and have become one of the most common ways in which corporate computers are breached.

Social media sites like Facebook, Instagram, Twitter, Reddit, and many others have changed how people relate to each other. Sometimes this is positive—keeping in touch with friends and family, watching news, entertainment of all sorts. Occasionally there are positive effects, for example viral videos of police brutality that brought Black Lives Matter to everyone's attention in mid-2020.

But social media have also contributed to a significant number of negatives. Racists, hate groups, conspiracy theorists and other crazy people, no matter what their beliefs or political positions, can easily find each other on the Internet, to coordinate and amplify their effects. Thorny questions about free speech and technological challenges in moderating content make it difficult to slow the spread of hatred and nonsense.

Jurisdictional issues are difficult in a world totally interconnected by the Internet. In 2018, the European Union implemented the General Data Protection Regulation (GDPR), which allows EU residents to control the collection and use of their personal data, and prevents companies from sending or storing such data outside the EU. The jury is still out on how effective the GDPR has been in improving individual privacy, and of course these rules apply only in the EU and are different in other parts of the world.

The rapid adoption of cloud computing, where individuals and companies store their data and do their computing on servers owned by Amazon, Google, Microsoft and the like, adds another layer of complexity. Data is no longer held directly by its owners but rather by third parties that have different agendas, responsibilities and vulnerabilities, and may face conflicting jurisdictional requirements.

There's a rapidly growing "Internet of Things" in which all kinds of devices connect to the Internet. Cell phones are an obvious instance, of course, but it's also cars, security cameras, home appliances and controls, medical equipment, and a great deal of infrastructure like air traffic control and power grids. This trend towards connecting everything to the Internet will continue, because the benefits of connection are compelling. Unfortunately, there are major risks, since some of these devices control life and death systems, not just our entertainment, and security for them is often much weaker than for more mature systems.

Cryptography is one of the few effective defenses against all of this, since it provides ways to keep communications and data storage private and secure. But strong cryptography is under continuous attack. Governments don't like the idea that individuals or companies or terrorists could have truly private communications, so there are frequent calls for putting backdoors into cryptographic algorithms to allow government agencies to break the encryption, though of course only with "proper safeguards" and only "in the interests of national security." However well-intentioned, this is a truly bad idea. Even if you believe that governments will always behave honorably and secret information will never leak (Snowden notwithstanding), weak cryptography helps your adversaries as well as your friends, and the bad guys won't use it anyway.

These are some of the problems and issues that ordinary people like the students in my course or the proverbial educated person on the street must worry about, no matter what their background and training.

The students in my course are not technical—no engineers, physicists or mathematicians. Instead they are English and politics majors, historians, classicists, economists, musicians and artists, a wonderful slice through the humanities and social sciences. By the end of the course these bright people should be able to read and understand a newspaper article about computing, to learn more from it, and perhaps to spot places where it might not be accurate. More broadly, I want my students and my readers to be intelligently skeptical about technology, to know that it is often a good thing but not a panacea. Conversely, though it sometimes has bad effects, technology is not an unmitigated evil.

A fine book by Richard Muller called *Physics for Future Presidents* attempts to explain the scientific and technical background underlying major issues that leaders have to grapple with—nuclear threats, terrorists, energy, global warming, and the like. Well-informed citizens without presidential aspirations should know something of these topics as well. Muller's approach is a good metaphor for what I would like to accomplish: "Computing for Future Presidents."

What should a future president know about computing? What should a well-informed person know about computing? What should *you* know?

I think there are four core technical areas: hardware, software, communications, and data.

Hardware is the tangible part, the computers we can see and touch, that sit in our homes and offices, and that we carry around in our phones. What's inside a computer, how does it work, how is it built? How does it store and process information?

What are bits and bytes, and how do we use them to represent music, movies, and everything else?

Software, the instructions that tell computers what to do, is by contrast hardly tangible at all. What can we compute, and how fast can we compute it? How do we tell computers what to do? Why is it so hard to make them work right? Why are they so often hard to use?

Communications means computers, phones, and other devices talking to each other on our behalf and letting us talk to each other: the Internet, the web, email and social networks. How do these work? The rewards are obvious, but what are the risks, especially to our privacy and security, and how can they be mitigated?

Data is all the information that hardware and software collect, store and process, and which communications systems send around the world. Some of this is data we contribute voluntarily, whether prudently or not, by uploading our words, pictures and videos. Much of it is personal information about us, gathered and shared without our knowledge, let alone agreement, as we go about our daily lives.

President or not, you should know about the world of computing because it affects you personally. No matter how non-technical your life and work, you're going to have to interact with technology and technical people. Knowing something of how devices and systems operate is a big advantage, even something as simple as recognizing when a salesperson, a help line or a politician is not telling you the whole truth.

Indeed, ignorance can be directly harmful. If you don't understand viruses, phishing and similar threats, you become more susceptible to them. If you don't know how social networks leak, or even broadcast, information that you thought was private, you're likely to reveal much more than you realize. If you're not aware of the headlong rush by commercial interests to exploit what they have learned about your life, you're giving up privacy for little benefit. If you don't know why it's risky to do your personal banking in a coffee shop or an airport, you're vulnerable to theft of money and identity. If you don't know how easily data can be manipulated, you're more likely to be taken in by fake news, fraudulent images, and conspiracy theories.

The book is meant to be read from front to back but you might prefer to skip ahead to topics of personal interest and come back later. For example, you could begin by reading about networks, cell phones, the Internet, the web and privacy issues starting in Chapter 8; you might have to look back at earlier chapters to understand a few parts, but mostly it will be accessible. You can skip anything quantitative, for instance how binary numbers work in Chapter 2, and ignore the details of programming languages in a couple of chapters.

The notes at the end list books that I like, and include links to sources and helpful supplements. The glossary gives brief definitions and explanations of important technical terms and acronyms.

Any book about computing can become dated quickly, and this one is no exception. The previous edition was published well before we learned about the extent to which hostile actors could sway public opinion and affect elections in the US and other countries. I've updated the book with important new stories, many of which relate to personal privacy and security, since that issue has become more pressing in the last few years. There's a new chapter on artificial intelligence, machine learning, and the role of "big data" in making them so effective and in some cases so dangerous. I've also tried to clarify explanations that were murky, and dated material has been deleted or replaced. Nevertheless, some details will be wrong or out of date when you read this, though I've tried to ensure that content of lasting value is clearly identified.

My goal for this book is that you will come away with some appreciation for an amazing technology and a real understanding of how it works, where it came from, and where it might be going in the future. Along the way, perhaps you'll pick up a helpful way of thinking about the world. I hope so.

Acknowledgments

I am again greatly indebted to friends and colleagues who have helped me to improve the book. As he has done so often, Jon Bentley read multiple drafts with great care; his organizational suggestions, fact-checking, and new examples have been invaluable. Al Aho, Swati Bhatt, Giovanni De Ferrari, Paul Kernighan, John Linderman, Madeleine Planeix-Crocker, Arnold Robbins, Yang Song, Howard Trickey and John Wait made detailed comments on the whole manuscript. I am also grateful to Fabrizio d'Amore, Peter Grabowski, Abigail Gupta, Maia Hamin, Gerard Holzmann, Ken Lambert, Daniel Lopresti, Theodor Marcu, Joann Ordille, Ayushi Sinha, William Ughetta, Peter Weinberger and Francisca Weirich-Freiberg for valuable suggestions. Sungchang Ha's Korean translation of the previous edition markedly improved this English version as well. Harry Lewis, John MacCormick, Bryan Respass and Eric Schmidt were generous with praise for the previous edition. As always, the production team at Princeton University Press—Mark Bellis, Lorraine Doneker, Kristen Hop, Dimitri Karetnikov and Hallie Stebbins—has been a pleasure to work with. MaryEllen Oliver's proofreading and fact-checking were exceptionally meticulous.

After twenty years, the students in my class are starting to run the world, or at least to help to keep it on the rails—journalists, doctors, lawyers, teachers at all levels, government officials, company founders, artists, performers, and deeply involved citizens. I'm very proud of them.

We are all indebted to the many people whose hard work and sacrifice during the Covid-19 crisis have made it possible for the rest of us to work in the comparative comfort of our homes, able to rely on essential services that kept functioning and medical systems that cared for us during the pandemic. No words can properly express how much we owe to them.

Acknowledgments for the first edition of *Understanding the Digital World*

I am again deeply indebted to friends and colleagues for their generous help and advice. As he did with the first edition, Jon Bentley read several drafts with meticulous care, providing helpful comments on every page; the book is much the better for his contributions. I also received valuable suggestions, criticisms and corrections on the whole manuscript from Swati Bhatt, Giovanni De Ferrari, Peter Grabowski, Gerard Holzmann, Vickie Kearn, Paul Kernighan, Eren Kursun, David Malan, David Mauskop, Deepa Muralidhar, Madeleine Planeix-Crocker, Arnold Robbins, Howard Trickey, Janet Vertesi and John Wait. I have also benefited from helpful advice from

David Dobkin, Alan Donovan, Andrew Judkis, Mark Kernighan, Elizabeth Linder, Jacqueline Mislow, Arvind Narayanan, Jonah Sinowitz, Peter Weinberger and Tony Wirth. The production team at Princeton University Press—Mark Bellis, Lorraine Doneker, Dimitri Karetnikov and Vickie Kearn—has been a pleasure to work with. My thanks to all of them.

I am also grateful to Princeton's Center for Information Technology Policy for good company, conversation, and weekly free lunches. And to the wonderful students of COS 109, whose talent and enthusiasm continue to amaze and inspire me, thank you.

Acknowledgments for *D is for Digital*

I am deeply indebted to friends and colleagues for generous help and advice. In particular, Jon Bentley provided detailed comments on almost every page of several drafts. Clay Bavor, Dan Bentley, Hildo Biersma, Stu Feldman, Gerard Holzmann, Joshua Katz, Mark Kernighan, Meg Kernighan, Paul Kernighan, David Malan, Tali Moreshet, Jon Riecke, Mike Shih, Bjarne Stroustrup, Howard Trickey, and John Wait read complete drafts with great care, made many helpful suggestions, and saved me from some major gaffes. I also thank Jennifer Chen, Doug Clark, Steve Elgersma, Avi Flamholz, Henry Leitner, Michael Li, Hugh Lynch, Patrick McCormick, Jacqueline Mislow, Jonathan Rochelle, Corey Thompson, and Chris Van Wyk for valuable comments. I hope that they will recognize the many places where I took their advice, but not notice the few where I did not.

David Brailsford offered a great deal of helpful advice on self-publishing and text formatting, based on his own hard-won experience. Greg Doench and Greg Wilson were generous with advice about publishing. I am indebted to Gerard Holzmann and John Wait for photographs.

Harry Lewis was my host at Harvard during the 2010–2011 academic year, when the first few drafts of the book were written. Harry's advice and his experience teaching an analogous course were of great value, as were his comments on multiple drafts. Harvard's School of Engineering and Applied Sciences and the Berkman Center for Internet and Society provided office space and facilities, a friendly and stimulating environment, and (yes, there is such a thing!) regular free lunches.

I am especially grateful to the many hundreds of students who have taken COS 109, "Computers in Our World." Their interest, enthusiasm and friendship have been a continual source of inspiration. I hope that when they are running the world a few years from now, they will have profited in some way from the course.

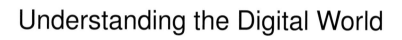

Understanding the Digital World

Introduction

"It was the best of times, it was the worst of times."

Charles Dickens, *A Tale of Two Cities*, 1859.

My wife and I had planned to spend the summer of 2020 on vacation in England. We made reservations, paid deposits, bought tickets, arranged with friends to look after our house and cat, and then the world changed.

By early March, it was clear that Covid-19 was going to be a major worldwide health crisis. Princeton University shut down physical classes and sent most students home on short notice. They were given a week to pack their belongings and leave, and it was quickly decided that they were not going to return that semester.

Classes moved online. Students watched lectures, wrote papers, took exams, and got their grades, all remotely. I became, if not an expert, at least an experienced amateur user of the Zoom video conferencing system. Fortunately, I was teaching two small seminars, with less than a dozen students in each, so it was possible to see everyone in the group at the same time, and to have reasonable conversations. Things were not so good for colleagues who were teaching large lecture classes, however, and of course the students on the other side of all these virtual lecterns were adversely affected.

Most students moved back to comfortable homes with reliable electric power, good Internet connectivity, a supportive family environment, and no shortages of food or other crucial supplies. Naturally, relationships suffered because of enforced separation, or prospered because of enforced togetherness, and sometimes vice versa. But these were minor problems.

Other students were in far worse situations. Some had intermittent or non-existent Internet connections that made video and email unusable. Some were ill or quarantined for extended periods. Some had sick relatives to care for, or even deaths in their families.

Day-to-day university administrative chores moved online as well, with casual hallway conversations converted into daily virtual meetings; paperwork was mostly replaced by email. Zoom fatigue set in quickly, but so far I have not been a victim of

Zoom bombing where some hacker invades my online space.

In many parts of the world, fortunate people were able to do their jobs online, and companies quickly shifted to "work from home" mode. People refined their video backdrops to show arrays of books, or neat displays of flowers and pictures, and they learned how to keep children, pets and significant others (mostly) quiet and out of the frame.

Streaming video from sources like Netflix, already popular, became more so. Online gaming grew as well, along with fantasy sports when real sports were canceled entirely.

We got continuously updated status reports on the rapid spread and discouragingly slow and erratic containment of Covid-19, albeit with far too much magical thinking and outright lies from politicians; honest and competent leaders were few and far between. We learned a bit about how quickly exponential processes grow.

It was surprisingly easy to adapt to this new way of doing business. The lucky ones were able to keep working, stay in virtual contact with friends and family, order food and supplies, almost as before. The Internet and all the infrastructure kept us connected. It was remarkably resilient—communications systems were always there, as, fortunately, were power, heat and water.

These technological systems worked so well during a global crisis that, aside from occasional anxious moments, we tended not to think about them, even though we would have been dead in the water without them, and, unsaid but so true, without the many heroic people behind the scenes who kept things working, often at grave risk to their own health and even lives. We also didn't think enough about the many millions of people who became unemployed, because their jobs couldn't be done via the Internet and just disappeared overnight.

I had literally never heard of Zoom before I had to start using it in March. Zoom was launched in 2013 to provide a video conferencing system that competed with larger operations like Microsoft Teams and Google's oft-renamed Meet. Zoom went public in 2019, and as I write this late in the fall of 2020, it is valued at over $125 billion, far more than older and better-known companies like General Motors ($61 billion) and General Electric ($85 billion), and well ahead of IBM ($116 billion).

Moving online worked for those who had fast, reliable Internet, and a computer with a camera and a microphone. Internet and cloud service providers had enough capacity to handle increased traffic. Video meeting services were commonplace and refined enough that most people were comfortable using them. None of this would have worked nearly so well, if at all, ten years earlier.

In short, ubiquitous modern technology made it possible for the fortunate to carry on a reasonable simulation of normality. This experience reminds us of the range of technology, how deeply it has become part of our lives, and how it has improved life in all kinds of ways.

But there's another side to the story, not so optimistic.

The Internet, already a hotbed of paranoia, hatred and crackpot theories, grew worse. Social media enabled politicians and government officials to spread lies, divide us further, and avoid blame, abetted by "news" outlets with no regard for facts. Sites like Twitter and Facebook tried unsuccessfully to find middle ground between being neutral platforms for the free expression of ideas and limiting the barrage of incendiary postings and outright falsehoods.

Surveillance has reached new heights, with technology used in many countries to restrict people and monitor and control their behavior. For instance, China uses face recognition for, among other things, keeping track of its minority populations. During the Covid-19 pandemic, the Chinese government mandated installation of an app that works as a sort of immunity passport but also reports its users' locations to the police. In the US and the UK, local law enforcement agencies use face recognition, license plate readers, and the like to keep an eye on people.

Our mobile phones continuously monitor our locations and a variety of parties are able to aggregate the data. Tracking applications for smartphones are an excellent example of the double-sided nature of technology. Who could be against a Covid-19 contact-tracing system that tells you whether you have been exposed to a potentially contagious person? But any technology that enables the government to know where you've been and who you've been talking to also helps them to monitor and control more effectively. It's a short and slippery slope from disease tracking to ferreting out peaceful protesters, dissidents, political enemies, whistle-blowers, and anyone else that the authorities think might be a threat. (It's not clear whether app-based contact tracing works at all, since it's susceptible to high false positive and false negative rates.)

For almost all of our interactions with the online world and often in the real world as well, countless computer systems watch and remember who you and I dealt with, how much we paid, and where we were at the time. A large part of this data gathering is for commercial purposes, since the more that companies know about us, the more accurately they can target us for advertising. Most readers know that such data is collected, but I expect that many would be surprised by how much there is and how detailed.

Companies are not the only observers; governments are deeply involved in surveillance as well. The NSA emails, internal reports, and PowerPoint presentations disclosed by Edward Snowden revealed much about spying in the digital era. The gist is that the NSA watches everyone on a grand scale.

Snowden's revelations were stunning. It had been widely suspected that the NSA spied on more people than it admitted, but the extent surpassed everyone's imagination. The NSA routinely collected metadata about all telephone calls made in the US—who called who, when they talked, and for how long—and may have recorded the content of these calls as well. It recorded my Skype conversations and email contacts, and probably the mail contents as well. (Yours too, of course.) It tapped the cell phones of world leaders. It intercepted huge amounts of Internet traffic by placing recording devices on equipment where submarine cables enter and leave the US. It enlisted or coerced the major telecommunications and Internet companies to gather and hand over information about their users. It stored great amounts of data for extended periods of time, sharing some of it with spy agencies in other countries.

Meanwhile, back on the commercial front, hardly a day goes by when we don't learn of another breach of security at some company or institution, in which shadowy hackers steal information like names, addresses, credit card numbers, and other personal information about millions of people. Usually these are high-tech criminals, but sometimes it's espionage by other countries, looking for valuable information. From time to time, foolish or careless behavior by whoever maintains the information accidentally exposes private data. No matter what the mechanism, data that has been collected about us is all too often exposed or stolen, potentially to be used against us.

The purpose of this book is to explain the technology that lies behind all of this, so you understand how such systems operate. How can pictures, music, movies, and intimate details of your personal life be sent around the world in no time at all? How do email and texting work, and how private are they? Why is spam so easy to send and so hard to get rid of? Do cell phones report where you are all the time? Who is tracking you online and on your phone, and why does that matter? Can your face be recognized in a crowd? Who knows that it's *your* face? Can hackers take over your car? How about self-driving cars? Can we defend our privacy and security, or should we just give up? By the end of the book, you should have a decent idea of how computer and communications systems work, how they affect you, and how you can strike a balance between using helpful services and protecting your privacy.

There are only a handful of fundamental ideas, which we will discuss in much more detail in the rest of the book.

First is the *universal digital representation of information*. Complex and sophisticated mechanical systems like those that stored documents, pictures, music and movies for much of the 20th century have been replaced by a single uniform storage mechanism. Information is represented digitally—as numeric values—rather than in specialized forms like colored dyes embedded in plastic film or magnetic patterns on vinyl tape. Paper mail gives way to digital mail. Paper maps yield to digital ones. Paper documents are replaced by online databases. All those disparate analog representations have been replaced by a common low-level representation in which everything is just numbers: digital information.

Second is the *universal digital processor*. All of this digital information can be processed by a single general-purpose device, the digital computer. Digital computers that process the uniform digital representation have replaced the elaborate and complicated mechanical devices that process analog representations. As we'll see, computers are all equally capable in what they can compute, differing only in how fast they operate and how much data they can store. A smartphone is a computer of great sophistication, with as much computing power as a laptop. Thus more and more of what might once have been limited to desktop or laptop computers has found its way onto phones, and this process of convergence is accelerating.

Third is the *universal digital network*. The Internet connects the digital computers that process the digital representation; it connects computers and phones to mail, search, social networks, shopping, banking, news, entertainment, and everything else. The majority of the world's population has access to this network. You can exchange email with anyone, regardless of where they might be or how they choose to access their mail. You can search, comparison shop, and purchase from your phone, laptop, or tablet. Social networks keep you in touch with friends and family, again from your phone or computer. You can watch endless entertainment, often for free. "Smart" devices monitor and control systems in your home; you can talk to them to tell them what to do or to ask them questions. There's a worldwide infrastructure that makes all these services work together.

Fourth, an immense amount of *digital data* is continuously being collected and analyzed. Maps, aerial photographs, and street-level views of much of the world are freely available. Search engines tirelessly scan the Internet so they can answer queries efficiently. Millions of books are available in digital form. Social networks and sharing sites maintain enormous amounts of data for and about us. Both online and brick-and-mortar stores and services provide access to their wares while quietly recording everything we do when we visit them, aided and abetted by search engines, social networks and our phones. For all of our online interactions, Internet service providers log the connections we make, and perhaps more. Governments spy on us all the time, to an extent and with a precision that would have been impossible a decade or two ago.

All of this is changing rapidly because digital technological systems continue to get smaller, faster, and cheaper. New phones with fancier features, better screens, and more interesting applications arrive continuously. New gadgets appear all the time; the most useful find their functionality subsumed into phone apps. This is a natural by-product of digital technology, in which any technological development leads to improvement across the board for digital devices: if some change makes it possible to handle data cheaper, faster or in larger quantity, all devices will benefit. As a result, digital systems are pervasive, an integral part of our lives both visibly and behind the scenes.

This progress must surely be a good thing, and indeed in most ways it is. But there are clouds around the silver lining. One of the most obvious and perhaps the most worrying to individuals is the impact of technology on personal privacy. When you use your phone to search for some product and then visit store web sites, all parties keep records of what you visited and what you clicked on. They know who you are because your phone identifies you uniquely. They know where you are because your phone reports its location to within a hundred meters or so all the time. The phone company records this information and may sell it. With GPS, the Global Positioning System, you can be located to within five to ten meters; with location services turned on, that information is available to apps, and they too can sell that information. In fact, it's even worse: disabling location services only prevents apps from using GPS data; it does not prevent the phone's operating system from collecting and uploading the data, which it could do by the cell network, Wi-Fi or Bluetooth.

You're being watched in real life as well as online. Face recognition technology can identify you on the street or in a store. Traffic cameras scan your license plates and know where your car is; so do electronic toll-collection systems. Internet-connected smart thermostats, voice responders, door locks, baby monitors and security cameras are surveillance devices that we have invited into our homes. The tracking that we permit today without even thinking about it makes the monitoring in George Orwell's *1984* look casual and superficial.

The records of what we do and where we do it may well live forever. Digital storage is so cheap and data is so valuable that information is rarely discarded. If you post something embarrassing online or send mail that you subsequently regret, it's too late. Information about you can be combined from multiple sources to create a detailed picture of your life, and is available to commercial, government and criminal interests without your knowledge or permission. It is likely to remain available indefinitely and could surface to harm you at any time in the future.

The universal network and its universal digital information have made us vulnerable to strangers to a degree never imagined a decade or two ago. As Bruce Schneier says in his excellent book *Data and Goliath*, "Our privacy is under assault from constant surveillance. Understanding how this occurs is critical to understanding what's at stake."

The societal mechanisms that protect our privacy and our property have not kept up with the rapid advances in technology. Thirty years ago, I dealt with my local bank and other financial institutions by physical mail and occasional personal visits. Accessing my money took time and it left an extensive paper trail; it would have been difficult for someone to steal from me. Today, I deal with financial institutions mostly through the Internet. I can conveniently access my data, but it's possible that through some blunder on my part or a screwup by one of these companies, someone on the far side of the world could clean out my account, steal my identity, ruin my credit rating, and who knows what else, in no time at all and with little chance of recourse.

This book is about understanding how these systems work and how they are changing our lives. Of necessity it's a snapshot, so you can be certain that ten years from now, today's systems will seem clunky and antiquated. Technological change is not an isolated event but an ongoing process—rapid, continuous, and accelerating. Fortunately, the basic ideas of digital systems will remain the same, so if you understand those, you'll understand tomorrow's systems too, and you'll be in a better position to deal with the challenges and the opportunities that they present.

Part I

Hardware

"I wish to God these calculations had been executed by steam."

Charles Babbage, 1821, quoted in Harry Wilmot Buxton,
Memoir of the Life and Labours of the Late Charles Babbage, 1872.

Hardware is the solid, visible part of computing: devices and equipment that you can see and put your hands on. The history of computing devices is interesting, but I will only mention a little of it here. Some trends are worth noting, however, especially the exponential increase in how much circuitry and how many devices can be packed into a given amount of space, often for a fixed price. As digital equipment has become cheaper and more powerful, widely disparate mechanical systems have been superseded by much more uniform electronic ones.

Computing machinery has a long history, though most early computational devices were specialized, often for predicting astronomical events and positions. For example, one (unproven) theory holds that Stonehenge was an astronomical observatory. The Antikythera mechanism, from about 100 BCE, is an astronomical computer of remarkable mechanical sophistication. Arithmetic devices like the abacus have been used for millennia, especially in Asia. The slide rule was invented in the early 1600s, not long after John Napier's description of logarithms. I used one as an undergraduate engineer in the 1960s, but slide rules are now curiosities, replaced by calculators and computers, and my painfully acquired expertise is useless.

The most relevant precursor to today's computers is the Jacquard loom, which was invented in France by Joseph Marie Jacquard around 1800. The Jacquard loom used rectangular cards with multiple rows of holes that specified weaving patterns. The Jacquard loom thus could be "programmed" to weave a wide variety of different patterns under the control of instructions that were provided on punched cards; changing the cards caused a different pattern to be woven. The creation of labor-saving machines for weaving led to social disruption as weavers were put out of work; the Luddite movement in England in 1811–1816 was a violent protest against mechanization. Modern computing technology has similarly led to disruption.

Figure I.1: Modern implementation of Babbage's Difference Engine.

Computing in today's sense began in England in the mid-19th century with the work of Charles Babbage. Babbage was a scientist who was interested in navigation and astronomy, both of which required tables of numeric values for computing positions. Babbage spent much of his life trying to build computing devices that would mechanize the tedious and error-prone manual arithmetic calculations needed to create the tables, and even to print them. You can sense his exasperation in the quotation on the previous page. For a variety of reasons, including alienating his financial backers, he never succeeded in his ambitions, but his designs were sound. Modern implementations of some of his machines, built with tools and materials from his time, can be seen in the Science Museum in London and the Computer History Museum in Mountain View, California (in the figure above).

Babbage encouraged a young woman, Augusta Ada Byron, the daughter of the poet George Byron, and later Countess of Lovelace, in her interests in mathematics and his computational devices. Lovelace wrote detailed descriptions of how to use Babbage's Analytical Engine (the most advanced of his planned devices) for scientific computation and speculated that machines could do non-numeric computation as well, such as composing music. "Supposing, for instance, that the fundamental relations of pitched sounds in the science of harmony and of musical composition were susceptible of such expression and adaptations, the engine might compose elaborate and scientific pieces of music of any degree of complexity or extent." Ada Lovelace is often called the world's first programmer, and the Ada programming language is

Figure I.2: Ada Lovelace. Detail from 1836 portrait by Margaret Sarah Carpenter.

named in her honor.

Herman Hollerith, working with the US Census Bureau in the late 1800s, designed and built machines that could tabulate census information far more rapidly than could be done by hand. Using ideas from the Jacquard loom, Hollerith used holes punched in stiff paper cards to encode census data in a form that could be processed by his machines. Famously, the 1880 census had taken eight years to tabulate, but with Hollerith's punch cards and tabulating machines, the 1890 census took only one year to prepare, instead of the predicted 10 years or more. Hollerith founded a company that in 1924 became, through mergers and acquisitions, International Business Machines, which we know today as IBM.

Babbage's machines were complex mechanical assemblies of gears, wheels, levers and rods. The development of electronics in the 20th century made it possible to imagine computers that did not rely on mechanical components. The first significant one of these all-electronic machines was ENIAC, the Electronic Numerical Integrator and Computer, which was built during the 1940s at the University of Pennsylvania in Philadelphia, by Presper Eckert and John Mauchly. ENIAC occupied a large room and required a large amount of electric power; it could do about 5,000 additions in a second. It was intended to be used for ballistics computations and the like, but it was not completed until 1946, well after the end of World War II. (Parts of ENIAC are on display in the Moore School of Engineering at the University of Pennsylvania.)

Babbage saw clearly that a computing device could store its operating instructions and its data in the same form, but ENIAC did not store instructions in memory along with data; instead it was programmed by setting up connections through switches and re-cabling. The first computers that truly stored programs and data together were built in England, notably EDSAC, the Electronic Delay Storage Automatic Calculator, at Cambridge in 1949.

Early electronic computers used vacuum tubes as computing elements. Vacuum tubes are electronic devices roughly the size and shape of a cylindrical light bulb (see Figure 1.7 in the next chapter); they were expensive, fragile, bulky, and power hungry. With the invention of the transistor in 1947, and then of integrated circuits in 1958, the modern era of computing really began. These technologies have allowed electronic systems to steadily become smaller, cheaper and faster.

The next three chapters describe computer hardware, focusing on the logical architecture of computing systems more than on the physical details of how they are built. The architecture has been largely unchanged for decades, while the hardware has changed to an astonishing degree. The first chapter is an overview of the structure and components of a computer. The second chapter shows how computers represent information with bits, bytes and binary numbers. The third chapter explains how computers actually compute: how they process the bits and bytes to make things happen.

1

What Is a Computer?

"Inasmuch as the completed device will be a general-purpose computing machine it should contain certain main organs relating to arithmetic, memory-storage, control and connection with the human operator."

Arthur W. Burks, Herman H. Goldstine, John von Neumann, "Preliminary discussion of the logical design of an electronic computing instrument," 1946.

Let's begin our discussion of hardware with an overview of what a computer is. We can look at a computer from at least two viewpoints: the logical or functional organization—what the pieces are, what they do and how they are connected—and the physical structure—what the pieces look like and how they are built. The goal of this chapter is to understand what a computer is, see what's inside, learn what each part does, and get a sense of what the myriad acronyms and numbers mean.

Think about your own computing devices. Many readers will have some kind of "PC," that is, a laptop or desktop computer descended from the Personal Computer that IBM first sold in 1981, running some version of the Windows operating system from Microsoft. Others will have an Apple Macintosh that runs a version of the macOS operating system. Still others might have a Chromebook running Chrome OS that relies on the Internet for most of its storage and computation. More specialized devices like smartphones, tablets and ebook readers are also powerful computers. These all look different and when you use them they feel different as well, but underneath the skin, they are fundamentally the same. We'll talk about why.

There's a loose analogy to cars. Functionally, cars have been the same for well over a hundred years. A car has an engine that uses some kind of fuel to make the engine run and the car move. It has a steering wheel that the driver uses to control the car. There are places to store the fuel and places to store the passengers and their goods. Physically, however, cars have changed greatly over a century: they are made of different materials, and they are faster, safer, and much more reliable and comfortable. There's a world of difference between my first car, a well-used 1959 Volkswagen Beetle, and a Ferrari, but either one will carry me and my groceries home from the store or across the country, and in that sense they are functionally the same.

(For the record, I have never even sat in a Ferrari, let alone owned one, so I'm specu-
lating about whether there's room for the groceries. I did park next to one once—
Figure 1.1.)

Figure 1.1:   The closest I've ever come to a Ferrari.

The same is true of computers. Logically, today's computers are very similar to
those of the 1950s, but the physical differences go far beyond the kinds of changes
that have occurred with the automobile. Today's computers are much smaller,
cheaper, faster and more reliable than those of 60 or 70 years ago, literally a million
times better in some properties. Such improvements are the fundamental reason why
computers are so pervasive.

The distinction between the functional behavior of something and its physical
properties—the difference between what it does and how it's built or works inside—
is an important idea. For computers, the "how it's built" part changes at an amazing
rate, as does how fast it runs, but the "what it does" part is quite stable. This distinction between an abstract description and a concrete implementation will come up
repeatedly in what follows.

I sometimes do a survey in my class in the first lecture. How many have a PC?
How many have a Mac? The ratio was fairly constant at 10 to 1 in favor of PCs in
the early 2000s, but changed rapidly over a few years, to the point where Macs now
account for well over three quarters of the computers. This is not typical of the world
at large, however, where PCs dominate by a wide margin.

Is the ratio unbalanced because one is superior to the other? If so, what changed
so dramatically in such a short time? I ask my students which kind is better, and for
objective criteria on which to base that opinion. What led you to your choice when
you bought your computer?

Naturally, price is one answer. PCs tend to be cheaper, the result of fierce competition in a marketplace with many suppliers. A wider range of hardware add-ons, more software, and more expertise are all readily available. This is an example of what economists call a *network effect*: the more other people use something, the more useful it will be for you, roughly in proportion to how many others there are.

On the Mac side are perceived reliability, quality, esthetics, and a sense that "things just work," for which many consumers are willing to pay a premium.

The debate goes on, with neither side convincing the other, but it raises some good questions and helps to get people thinking about what is different between various kinds of computing devices and what is really the same.

There's an analogous debate about phones. Almost everyone has a "smartphone" that can run programs ("apps") downloaded from Apple's App Store or the Google Play Store. The phone serves as a browser, a mail system, a watch, a camera, a music and video player, a voice recorder, a map, a navigator, a comparison shopping tool, and even occasionally a device for conversation. Typically about three quarters of my students have an iPhone; almost all the rest have an Android phone from one of many suppliers. iPhones are more expensive but offer smooth integration with Apple's ecosystem of computers, tablets, watches, music players, and cloud services, another example of a network effect. Rarely, someone admits to having only a "feature phone," which is defined as a phone that has no features beyond the ability to make phone calls. My sample is for the US and a comparatively affluent environment; in other environments and other parts of the world, Android phones would be much more common.

Again, people have good reasons—functional, economic, esthetic—for choosing one kind of phone over others but underneath, just as for PCs versus Macs, the hardware that does the computing is very similar. Let's look at why.

1.1 Logical Construction

If we were to draw an abstract picture of a simple generic computer—its logical or functional architecture—it would look like the diagram in Figure 1.2 for both a Mac and a PC: a processor, some primary memory, some secondary storage, and a variety of other components, all connected by a set of wires called a *bus* that carries information between them.

If instead we drew this picture for a phone or tablet, it would be similar, though mouse, keyboard and display are combined into one component, the screen, and there are many hidden components like a compass, an accelerometer, and a GPS receiver for determining your physical location.

The basic organization—a processor, memory and storage for instructions and data, and input and output devices—has been standard since the 1940s. It's often called the *von Neumann architecture*, after John von Neumann, who described it in the 1946 paper quoted above. Though there is still occasional debate over whether von Neumann gets too much credit for work done by others, the paper is so clear and insightful that it is well worth reading even today. For example, the quotation at the beginning of this chapter is the first sentence of the paper. Translated into today's terminology, the processor provides arithmetic and control, the primary memory and

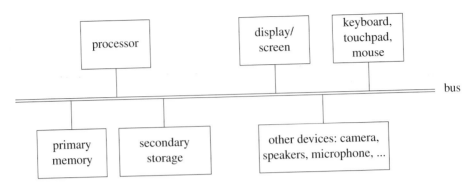

Figure 1.2: Architectural diagram of a simple idealized computer.

secondary storage are memory-storage, and the keyboard, mouse and display interact with the human operator.

A note on terminology: the processor has historically been called the *CPU* or *central processing unit*, but is now more often just "processor." The primary memory is often called *RAM* or *random access memory*, and secondary storage is often *disk* or *drive*, reflecting different physical implementations. I'll mostly use the words processor, memory and storage, but with occasional lapses into the older terms.

1.1.1 Processor

The processor is the brain, if a computer could be said to have such a thing. The processor does arithmetic, moves data around, and controls the operation of the other components. The processor has a limited repertoire of basic operations that it can perform but it does them blazingly fast, billions per second. It can decide what operations to do next based on the results of previous computations, so it is to a considerable degree independent of its human users. We will spend more time on this component in Chapter 3 because it's so important.

If you go to a store or shop online to buy a computer, you'll find most of these components mentioned, usually accompanied by mysterious acronyms and equally mysterious numbers. For example, you might see a processor described as a "2.2 GHz dual-core Intel Core i7," as it is for one of my computers. What's that? Intel makes the processor and "Core i7" is Intel's name for an extensive line of processors. This particular processor has two processing units in a single package; in this context, lower case "core" has become a synonym for "processor." A core is a processor in its own right, but the CPU may have several cores that can work together or independently to compute faster. For most purposes, it's sufficient to think of the combination as "the processor," no matter how many cores it has.

"2.2 GHz" is the more interesting part. Processor speed is measured, at least approximately, in terms of the number of operations or instructions or parts thereof that it can do in a second. The processor uses an internal clock, rather like a heartbeat or the ticking of a clock, to step through its basic operations. One measure of speed is the number of such ticks per second. One beat or tick per second is called one *hertz* (abbreviated *Hz*), after the German engineer Heinrich Hertz, whose 1888 discovery of how to produce electromagnetic radiation led directly to radio and other wireless systems. Radio stations give their broadcast frequencies in megahertz (millions of hertz), like 102.3 MHz. Computers today typically run in the billions of hertz, or gigahertz, or GHz; my quite ordinary 2.2 GHz processor is zipping along at 2,200,000,000 ticks per second. The human heartbeat is about 1 Hz or almost 100,000 beats per day, which is around 30 million per year. So each core in my processor does in 1 second the number of beats my heart would do in 70 years.

This is our first encounter with some of the numerical prefixes like mega and giga that are so common in computing. "Mega" is one million, or 10^6; "giga" is one billion, or 10^9, and pronounced with a hard "g" as in "gig." We'll see more units soon enough, and there is a complete table in the glossary.

1.1.2 Primary memory

The primary memory stores information that is in active use by the processor and other parts of the computer; its contents can be changed by the processor. The primary memory stores not only the data that the processor is currently working on, but also the instructions that tell the processor what to do with that data. This is a crucially important point: by loading different instructions into memory, we can make the processor do a different computation. This makes the stored-program computer a general-purpose device; the same computer can run a word processor and a spreadsheet, surf the web, send and receive email, keep up with friends on Facebook, do my taxes, and play music, all by placing suitable instructions in the memory. The importance of the stored-program idea cannot be overstated.

The primary memory provides a place to store information while the computer is running. It stores the instructions of programs that are currently active, like Word, Photoshop or a browser. It stores their data—the documents being edited, the pictures on the screen, the music that's currently playing. It also stores the instructions of the operating system—Windows, macOS or something else—that operates behind the scenes to let you run multiple applications at the same time. We'll talk about applications and operating systems in Chapter 6.

The primary memory is called random access memory or RAM because the processor can access the information stored at any place within it as quickly as in any other; to over-simplify a little, there's no speed penalty for accessing memory locations in a random order. Though they passed from the scene long ago, you might remember VCR tapes, where to look at the end of a movie, you had to fast forward (slowly!) over everything from the beginning; that's called *sequential access*.

Most RAM is *volatile*, that is, its contents disappear if the power is turned off, and all of the currently active information is lost. That's why it's prudent to save your work often, especially on a desktop machine, where tripping over the power cord could be a real disaster.

Your computer has a fixed amount of primary memory. Capacity is measured in bytes, where a *byte* is an amount of memory that's big enough to hold a single character like W or @, or a small number like 42, or a part of a larger value. Chapter 2 will show how information is represented in memory and other parts of a computer, since it's one of the fundamental issues in computing. But for now, you can think of the memory as a large collection of identical little boxes, numbered up to a few billion, each of which can hold a small amount of information.

What is the capacity? The laptop I'm using right now has 8 billion bytes or 8 gigabytes or 8 GB of primary memory, which is perhaps too small. The reason is that more memory usually translates into faster computing, since there's never enough for all the programs that want to use it at the same time, and it takes time to move parts of an inactive program out to make room for something new. If you want your computer to run faster, buying extra RAM is likely to be the best strategy, at least if the memory is upgradable—it might not be.

1.1.3 Secondary storage

The primary memory has a large but limited capacity to store information; its contents disappear when the power is turned off. *Secondary storage* holds information even when the power is off. There are two main kinds of secondary storage: the older magnetic disk, called the *hard disk* or *hard drive*, and the newer form called the *solid state drive* or *SSD*. Both kinds of drive store much more information than the primary memory and it's not volatile: information on either kind of drive stays there even if there is no power. Data, instructions, and everything else is stored on secondary storage for the long term and brought into primary memory only transiently.

Magnetic disks store information by setting the direction of magnetization of tiny regions of magnetic material on rotating metallic surfaces. Data is stored in concentric tracks that are read and written by a sensor that moves from track to track. The whirring and clicking that you heard when an older computer was doing something is the disk in action, moving the sensor to the right places on the surface. The disk surface rotates at high speed, at least 5,400 revolutions per minute. You can see the surface and sensor in the picture of a standard laptop disk in Figure 1.3; the platter is 2.5 inches (6.35 cm) in diameter.

Disk storage is about 100 times cheaper per byte than RAM, but accessing information is slower. It takes about ten milliseconds for the disk drive to access any particular track on the surface; data is then transferred at roughly 100 MB per second.

Ten years ago, almost all laptops had magnetic disks. Today almost all have SSD, which uses *flash memory* instead of rotating machinery. Flash memory is non-volatile; information is stored as electric charges in circuitry that maintains the charge in individual circuit elements even when the power is off. Stored charges can be read to see what their values are, and they can be erased and overwritten with new values. Flash memory is faster, lighter, more reliable, won't break if dropped, and requires less power than conventional disk storage, so it's also used in cell phones, cameras, and the like. It's still more expensive per byte but prices are coming down and the advantages are so compelling that SSD has pretty much taken over from mechanical disks in laptops.

A typical laptop SSD holds 250 to 500 gigabytes. External drives that can be plugged in to a USB socket have capacities in the multi-terabyte (TB) range; they are still based on rotating machinery. "Tera" is one trillion, or 10^{12}, another unit that you'll see often.

Figure 1.3: Inside a hard disk drive.

How big is a terabyte, or even a gigabyte for that matter? One byte holds one alphabetic character in the most common representation of English text. *Pride and Prejudice*, about 250 pages on paper, has about 680,000 characters, so 1 GB could hold nearly 1,500 copies of it. More likely, I would store one copy and then include some music. Music in MP3 format is about 1 MB per minute, so an MP3 version of one of my favorite audio CDs, *The Jane Austen Songbook*, is about 60 MB, and there would still be room for another 15 hours of music in one gigabyte. The two-disk DVD of the 1995 BBC production of *Pride and Prejudice* with Jennifer Ehle and Colin Firth is less than 10 GB, so I could store it and a hundred similar movies in one terabyte.

A disk drive is a good example of the difference between logical structure and physical implementation. Programs like File Explorer on Windows or Finder on macOS display the contents of a drive as a hierarchy of folders and files. But the data could be stored on rotating machinery, integrated circuits with no moving parts, or something else entirely. The particular kind of drive in a computer doesn't matter. Hardware in the drive itself and software in the operating system, called the file system, work together to create the organizational structure. We will return to this in Chapter 6.

The logical organization is so well matched to people (or, more likely, by now we're so completely used to it) that other devices provide the same organization even though they use completely different physical means to achieve it. For example, the software that gives you access to information from a CD-ROM or DVD makes it look like this information is stored in a file hierarchy, regardless of how it is physically stored. So do USB devices, cameras and other gadgets that use removable memory cards. Even the venerable floppy disk, now totally obsolete, looked the same at the logical level. This is a good example of *abstraction*, a pervasive idea in computing: physical implementation details are hidden. In the file system case, no matter how the different technologies work, the contents are presented to users as a hierarchy of files and folders.

1.1.4 Et cetera

Myriad other devices serve special functions. Mice, keyboards, touch screens, microphones, cameras and scanners all allow users to provide input. Displays, printers and speakers output to users. Networking components like Wi-Fi or Bluetooth communicate with other computers. A variety of assistive technologies help people cope with vision, hearing or other access problems.

The architecture drawing in Figure 1.2 shows these as if they were all connected by a set of wires called a *bus*, a term borrowed from electrical engineering. In reality, there are multiple buses inside a computer, with properties appropriate to their function—short, fast, and expensive between processor and memory; long, slow, but cheap to the headphone jack. Some of the buses make an appearance outside as well, like the ubiquitous Universal Serial Bus or *USB* for plugging devices into a computer.

We won't spend much time on other devices at the moment, though we'll occasionally mention them in some specific context. For now, try to list the different devices that might accompany your computer or be attached to it: mice, keyboards, touchpads and touch screens, displays, printers, scanners, game controllers, headphones, speakers, microphones, cameras, phones, fingerprint sensors, connections to other computers. The list goes on and on. All of these have gone through the same evolution as processors, memory, and disk drives: the physical properties have changed rapidly, usually towards more capabilities in a smaller package at a lower price.

It's also worth noting how these devices are converging into a single one. Cell phones now serve as watches, calculators, still and video cameras, music and movie players, game consoles, barcode readers, navigators, and even flashlights. A smartphone has the same abstract architecture as a laptop, though with major implementation differences due to size and power constraints. Phones don't have hard disks like the one shown in Figure 1.3, but they do have flash memory to store information—contact lists, pictures, apps, and the like—when the phone is turned off. They don't have many external devices either, though there's likely a socket for headphones, and a USB connector. Tiny cameras are so cheap that most phones have one on each side. Tablets like the iPad and its competitors occupy another position in the space of possibilities; they too are computers with the same general architecture and similar components.

1.2 Physical Construction

In class, I pass around a variety of hardware devices (the legacy of decades of dumpster diving), with their innards exposed. So many things in computing are abstract that it's helpful to be able to see and touch disks, integrated circuit chips, the wafers on which they are manufactured, and so on. It's also interesting to see the evolution of some of these devices. For example, a laptop hard drive today is indistinguishable from one a decade or two old; the newer one is likely to have 10 or 100 times the capacity but the improvement is invisible. The same is true of Secure Digital (*SD*) cards like those used in digital cameras. Today's packages are identical to those of a few years ago (Figure 1.4), but the capacity is much higher and the price is lower; that 32 GB card costs less than 10 dollars.

Figure 1.4: SD cards of very different capacities.

On the other hand, there's a clear progression in the circuit boards that hold the components of a computer; there are fewer components today because more of the circuitry is inside them, the wiring is finer, and the connecting pins are more numerous and much closer together than they were 20 years ago.

Figure 1.5: PC circuit board, circa 1998; 12 x 7.5 inches (30 x 19 cm).

Figure 1.5 shows a desktop PC circuit board from the late 1990s. The components like the processor and the memory are mounted on or plugged into this board, and are connected by wires printed on the other side. Figure 1.6 shows part of the back side of the circuit board in Figure 1.5; the parallel printed wires are buses of various sorts.

Figure 1.6: Buses on printed circuit board.

Electronic circuits in computers are built from large numbers of a handful of basic elements. The most important of these is the *logic gate*, which computes a single output value based on one or two input values; it uses input signals like voltage or current to control an output signal, also voltage or current. Given enough of such gates connected in the right way, it's possible to perform any kind of computation. Charles Petzold's book *Code* is a nice introduction to this, and numerous web sites offer graphical animations that show how logic circuits can do arithmetic and other computations.

The fundamental circuit element is the *transistor*, a device invented at Bell Labs in 1947 by John Bardeen, Walter Brattain and William Shockley, who shared the 1956 Nobel Prize in physics for their invention. In a computer, a transistor is basically a switch, a device that can turn a current on or off under the control of a voltage; with this simple foundation, arbitrarily complicated systems can be constructed.

Logic gates used to be built from discrete components—vacuum tubes the size of light bulbs in ENIAC and individual transistors about the size of a pencil eraser in the computers of the 1960s. Figure 1.7 shows a replica of the first transistor (on the left), a vacuum tube, and a processor in its package; the actual circuit part is at the center and is about 1 cm square; the vacuum tube is about 4 inches (10 cm) long. A modern processor of this size would contain several billion transistors.

Logic gates are created on *integrated circuits* or ICs, often called *chips* or *microchips*. An integrated circuit has all the components and wiring of an electronic

Figure 1.7:   Vacuum tube, first transistor, processor chip in package.

circuit on a single flat surface (a thin sheet of silicon) that is manufactured by a complex sequence of optical and chemical processes to produce a circuit that has no discrete pieces and no conventional wires. ICs are thus much smaller and far more
robust than discrete-component circuitry. Chips are fabricated en masse on circular
wafers about 12 inches (30 cm) in diameter; the wafers are sliced into separate chips
that are individually packaged. A typical chip (Figure 1.7, bottom right) is mounted
in a larger package with dozens to hundreds of pins that connect it to the rest of the
system. Figure 1.8 shows an integrated circuit in its package; the actual processor is
at the center, and is about 1 cm square.

The fact that integrated circuits are based on silicon led to the nickname *Silicon
Valley* for the region in California south of San Francisco where the integrated circuit
business first took off; it's now a shorthand for all the high-tech businesses in the
area, and the inspiration for dozens of wannabes like Silicon Alley in New York and
Silicon Fen in Cambridge, England.

ICs were independently invented around 1958 by Robert Noyce and Jack Kilby;
Noyce died in 1990, but Kilby shared the 2000 Nobel Prize in physics for his role.
Integrated circuits are central to digital electronics, though other technologies are
used as well: magnetic storage for disks, lasers for CDs and DVDs, and optical fiber
for networking. All of these have had dramatic improvements in size, capacity and
cost over the past 50 or 60 years.

Figure 1.8:   Integrated circuit chip.

1.3 Moore's Law

In 1965, Gordon Moore, later the co-founder and long-time CEO of Intel, published a short article entitled "Cramming more components onto integrated circuits."
Extrapolating from a very few data points, Moore observed that as technology
improved, the number of transistors that could be manufactured on an integrated circuit of a particular size was doubling approximately every year, a rate that he later
revised to every two years, and others have set at 18 months. Since the number of
transistors is a rough surrogate for computing power, this meant that computing
power was doubling every two years, if not faster. In 20 years there would be ten
doublings and the number of devices would have increased by a factor of 2^{10}, that is,
about one thousand. In forty years, the factor is a million or more.

This exponential growth, now known as *Moore's Law*, has been going on for
nearly sixty years, so integrated circuits now have well over a million times as many
transistors as they did in 1965. Graphs of Moore's Law in action, especially for processor chips, show the number of transistors rising from a couple of thousand in
Intel's 8008 processor in the early 1970s to billions in the processors in inexpensive
consumer laptops today.

The single number that best characterizes the scale of circuitry is the size of individual features on an integrated circuit, for example, the width of a wire or the active
part of a transistor. This number has been shrinking steadily for many years. The
first (and only) integrated circuit I ever designed used 3.5 micron (3.5 micrometer)
features in 1980. For many ICs in 2021, the minimum feature size is 7 nanometers,
that is, 7 billionths of a meter, and the next step will be 5 nanometers. "Milli" is one
thousandth, or 10^{-3}; "micro" is one millionth, or 10^{-6}; "nano" is one billionth, or
10^{-9}, and nanometer is abbreviated nm. For comparison, a sheet of paper or a human
hair is about 100 micrometers or 1/10th of a millimeter thick.

If the width of features on an integrated circuit shrinks by a factor of 1,000, then the number of components in a given area increases by the square, that is, by a factor of a million. That factor is what takes a thousand transistors in an older technology into a billion in newer technology.

The design and manufacture of integrated circuits is an extremely sophisticated business, and highly competitive. Manufacturing operations ("fabrication lines") are expensive as well; a new factory can cost billions of dollars. A company that can't keep up technically and financially is at a competitive disadvantage, and a country that doesn't have such resources must depend on others for its technology, potentially a serious strategic problem.

Moore's Law is not a law of nature, but a guideline that the semiconductor industry has used to set targets. At some point the law will stop working. Its limits have often been predicted in the past, though ways around them have been found so far. We are getting to the point where there are only a handful of individual atoms in some circuits, however, and that's too small to control.

Processor speeds are not growing much, certainly no longer doubling every couple of years, in part because faster chips generate too much heat, but memory capacity still does increase. Meanwhile, processors can use more transistors by placing more than one processor core on a chip, and systems often have multiple processor chips; the growth is in the number of cores, not in how fast individual ones run.

It's striking to compare a personal computer of today to the original IBM PC, which appeared in 1981. That PC had a 4.77 MHz processor; the clock rate in a 2.2 GHz processor core is nearly 500 times faster, and there are likely two or four cores. It had 64 kilobytes of RAM; today's 8 GB computers have 125,000 times as much. ("Kilo" is one thousand, abbreviated "K".) The first PC had at most 750 KB of floppy disk storage and no hard disk; today's laptops are creeping up on a million times as much secondary storage. The PC had an 11-inch screen that could only display 24 rows of 80 green letters on a black background; I wrote much of this book on a 24-inch screen with 16 million colors. A PC with 64 KB of memory and a single 160 KB floppy disk cost $3,000 in 1981 dollars, which now might be equivalent to $10,000; today a laptop with a 2 GHz processor, 8 GB of RAM, and a 256 GB solid state drive costs a few hundred dollars.

1.4 Summary

Computer hardware, indeed digital hardware of all sorts, has been improving exponentially for sixty years, starting with the invention of the integrated circuit. The word "exponential" is often misunderstood and misused, but in this case it's accurate; over every fixed period of time, circuits have consistently gotten smaller or cheaper or more capable by a given percentage. The simplest version is Moore's Law: every 18 months or so the number of devices that can be put on an integrated circuit of a given size approximately doubles. This tremendous growth in capabilities is at the heart of the digital revolution that has changed our lives so much.

This growth in capability and capacity has also changed our notions of what computing and computers are. The first computers were viewed as number crunchers, suitable for ballistics, weapon design, and other scientific and engineering computations. The next use was business data processing—computing payrolls, generating invoices, and so on, and then as storage became cheaper, managing the databases that kept track of the information needed for computing those payrolls and bills. With the advent of the PC, computers became cheap enough that anyone could afford one and they began to be used for personal data processing, keeping track of home finances, and word processing tasks like writing letters. Not long after that, they also began to be used for entertainment: playing music CDs, and especially for games. And when the Internet appeared, our computers became communications devices as well, providing mail, the web, and social media.

The basic architecture of a computer—what the pieces are, what they do, and how they are connected to each other—has not changed since the 1940s. If von Neumann were to come back and examine one of today's computers, I conjecture that he would be stunned by the capabilities and the applications of modern hardware but he would find the architecture completely familiar.

Computers used to be physically huge, occupying large air-conditioned rooms, but they have shrunk steadily. Laptops today are about as small as they can get while remaining useful. The computers inside our phones are just as powerful, and phones are also about as small as they can reasonably be. The computers inside our gadgets are tiny too, as are the gadgets themselves in many cases. At the other end of the spectrum, we routinely deal with "computers" that live in data centers (back to the air-conditioned rooms) somewhere. We shop, search, and talk with friends using those computers, without even thinking of them as computers, let alone worrying about where they might be. They are just "there" somewhere in the cloud.

One of the great insights of 20th century computer science is that the logical or functional properties of today's digital computers, the original PC, its physically much bigger but less powerful ancestors, our ubiquitous phones, our computer-enabled devices, and the servers that provide cloud computing are all the same. If we ignore practicalities like speed and storage capacity, they all can compute exactly the same things. Thus, improvements in hardware make a great practical difference in what we can compute, but surprisingly do not of themselves make any fundamental change in what could be computed in principle. We'll talk more about this in Chapter 3.

2

Bits, Bytes, and Representation of Information

"If the base 2 is used the resulting units may be called binary digits, or more briefly *bits*, a word suggested by J. W. Tukey."

Claude Shannon, *A Mathematical Theory of Communication*, 1948.

In this chapter we're going to discuss three fundamental ideas about how computers represent information.

First, *computers are digital processors*: they store and process information that comes in discrete chunks and takes on discrete values—basically just numbers. By contrast, analog information implies smoothly varying values.

Second, *computers represent information in bits*. A *bit* is a binary digit, that is, a number that is either 0 or 1. Everything inside the computer is represented with bits instead of the familiar decimal numbers that people use.

Third, *groups of bits represent larger things*. Numbers, letters, words, names, sounds, pictures, movies, and the instructions that make up the programs that process them—all of these are represented as groups of bits.

You can safely skip the numeric details in this chapter, but the ideas are important.

2.1 Analog versus Digital

Let's distinguish between analog and digital. "Analog" comes from the same root as "analogous," and is meant to convey the idea of values that change smoothly as something else changes. Much of what we deal with in the real world is analog, like a water tap or the steering wheel of a car. If you want to turn the car a little, you turn the wheel a little; you can make as small an adjustment as you like. Compare this to the turn signal, which is either on or off; there's no middle ground. In an analog device, something (how much the car turns) varies smoothly and continuously in proportion to a change in something else (how much you turn the steering wheel). There are no discrete steps; a small change in one thing implies a small change in another.

Digital systems deal with discrete values, so there are only a fixed number of possible values: the turn signal is either off or it's on in one direction or the other. A small change in something results either in no change or in a sudden change in something else, from one of its discrete values to another.

Think about a watch. Analog watches have an hour hand, a minute hand, and a second hand that goes around once a minute. Although modern watches are controlled by digital circuitry inside, the hour and minute hands move smoothly through every possible position as time passes. By contrast, a digital watch or a cell phone clock displays time with digits. The display changes every second, a new minute value appears every minute, and there's never a fractional second.

Think about a car speedometer. My car has a traditional analog speedometer, where a needle moves smoothly in direct proportion to the car's speed. The transitions from one speed to another are smooth and there's no break. But it also has a digital display that shows speed to the nearest mile or kilometer per hour. Go a tiny bit faster and the display goes from 65 to 66; go a tiny bit slower and it drops back to 65. There's never a display of 65.5.

Think about a thermometer. The kind with a column of red liquid (colored alcohol, usually) or mercury is analog: the liquid expands or contracts in direct proportion to temperature changes, so a small change in temperature will produce a similarly small change in the height of the column. But the sign that flashes 37° outside a building is digital: the display is numeric, and it shows 37 for all temperatures between 36½ and 37½.

This can lead to some odd situations. Years ago, I was listening to my car radio on a US highway within reception distance of Canada, which uses the metric system. The announcer, trying to be helpful to everyone in his audience, said "the Fahrenheit temperature has gone up one degree in the last hour; the Celsius temperature is unchanged."

Why digital instead of analog? After all, our world is analog, and analog devices like watches and speedometers are easier to interpret at a glance. Nevertheless, much modern technology is digital; in many ways, that's the story told in this book. Data from the external world—sound, images, movement, temperature, and everything else—is converted as soon as possible to a digital form on the input side, and is converted back to analog form as late as possible on the output side. The reason is that digital data is easy for computers to work with. It can be stored, transported, and processed in many ways regardless of its original source. As we'll see in Chapter 8, digital information can be compressed by squeezing out redundant or unimportant information. It can be encrypted for security and privacy, it can be merged with other data, it can be copied exactly, it can be shipped anywhere via the Internet, and it can be stored in an endless variety of devices. Most of this is infeasible or even impossible with analog information.

Digital systems have another advantage over analog: they are much more easily extended. In stopwatch mode, my digital watch can display elapsed times to a hundredth of a second; adding that capability to an analog watch would be challenging. On the other hand, analog systems sometimes have the advantage: old media like clay tablets, stone carvings, parchment, paper and photographic film have all stood the test of time in a way that digital forms may not.

2.2 Analog-Digital Conversion

How do we convert analog information into digital form? Let's look at some of the basic examples, beginning with pictures and music, which between them illustrate the most important ideas.

2.2.1 Digitizing images

Conversion of images to digital form is probably the easiest way to visualize the process. Suppose we take a picture of the family cat (Figure 2.1).

Figure 2.1: The family cat in 2020.

An analog camera creates an image by exposing a light-sensitive area of chemical-coated plastic film to light from the object being photographed. Different areas receive different amounts of light of different colors, and that affects dyes in the film. The film is developed and printed on paper through a complicated sequence of chemical processes; the colors are displayed as varying amounts of colored dyes.

In a digital camera, the lens focuses the image onto a rectangular array of tiny light-sensitive detectors that lie behind red, green and blue filters. Each detector stores an amount of electric charge that is proportional to the amount of light that falls on it. These charges are converted into numeric values and the digital representation of the picture is the sequence of resulting numbers that represent the light intensities. If the detectors are more numerous and the charges are measured more precisely, then the digitized image will capture the original more accurately.

Each element of the sensor array is a trio of detectors that measure the amount of red, green and blue light; each group is called a *pixel*, for picture element. If the image is 4,000 by 3,000 pixels, that's twelve million picture elements, or twelve megapixels, small for today's digital cameras. The color of a pixel is usually represented by three values that record the intensities of red, green and blue that it contains, so a twelve megapixel image has 36 million light intensity values altogether. Screens display images on arrays of triplets of tiny red, green and blue lights whose brightness levels are set by the corresponding levels in the pixel; if you look at the screen of a phone, computer or TV with a magnifying glass, you can see the individual colored spots, somewhat like those in Figure 2.2. If you're close enough, you can see the same thing on stadium screens and digital billboards.

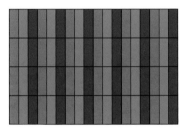

Figure 2.2: RGB pixels.

2.2.2 Digitizing sound

The second example of analog-to-digital conversion is sound, especially music. Digital music is a nice example because it was one of the first areas where the properties of digital information began to have major social, economic and legal implications. Unlike vinyl records or audio tape cassettes, digital music can be copied perfectly on any home computer as many times as desired, for free, and the perfect copies can be sent anywhere in the world without error via the Internet, also for free. The recorded music industry saw this as a serious threat and began a campaign of legal and political action to try to suppress copying. The war is not over—skirmishes are still fought in courts and political arenas—but the advent of streaming music services like Spotify has reduced the problem. We'll come back to this in Chapter 9.

What is sound? A sound source creates fluctuations in air pressure by vibration or other rapid motion, and our ears convert the pressure changes into neural activity that our brains interpret as sound. In the 1870s, Thomas Edison built a device that he called a "phonograph," which converted the fluctuations into a pattern of grooves in a wax cylinder that could be used later to recreate the air pressure fluctuations. Converting a sound into a pattern of grooves was "recording"; converting from the pattern to fluctuations in air pressure was "playback." Edison's invention was rapidly refined, and by the 1940s had evolved into the long-playing record or LP (Figure 2.3), which is still in use today, though primarily by retro sound enthusiasts.

LPs are vinyl disks with long spiral grooves that encode variations in sound pressure over time. A microphone measures variations in pressure as a sound is produced. These measurements are used to create a pattern on the spiral groove. When the LP is played, a fine needle follows the pattern in the groove and its motion is converted into a fluctuating electrical current that is amplified and used to drive a loudspeaker or an earphone, devices that create sound by vibrating a surface.

Figure 2.3: LP ("long-playing") record.

It's easy to visualize sound by plotting how air pressure changes with time, as in the graph in Figure 2.4. The pressure can be represented in any number of physical ways: voltage or current in an electronic circuit, brightness of a light, or a purely mechanical system as it was in Edison's original phonograph. The height of the sound pressure wave is the sound intensity or loudness, and the horizontal dimension is time; the number of waves per second is the pitch or frequency.

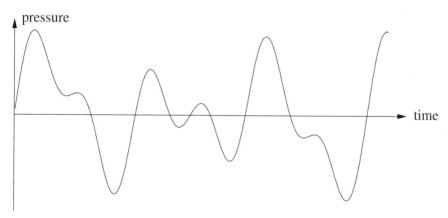

Figure 2.4: Sound waveform.

Suppose we measure the height of the curve—the air pressure at the microphone, perhaps—at regular intervals, as indicated by the vertical lines in Figure 2.5.

The measurements provide a sequence of numeric values that approximate the curve; the more often and the more precisely we measure, the more accurate the approximation will be. The resulting sequence of numbers is a *digital representation* of the waveform that can be stored, copied, manipulated and shipped elsewhere. We

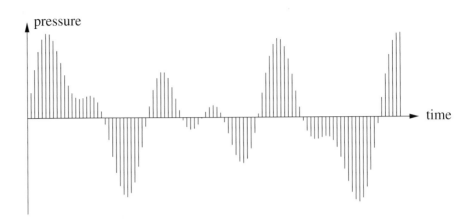

Figure 2.5: Sampling a sound waveform.

can do playback with a device that converts the numeric values into a matching pattern of voltage or current to drive a speaker or earphone and thus render it back into sound. Converting from the waveform to numbers is analog-to-digital conversion and the device is called an A/D converter; the other direction is of course digital-to-analog conversion, or D/A. Conversion is never perfect; something is lost in each direction. For most people the loss is imperceptible, though audiophiles claim that digital sound is not as good as LPs.

The audio compact disc or CD appeared around 1982, and was the first consumer example of digital sound. Rather than the analog groove of an LP record, a CD records *numbers* in a long spiral track on one side of the disk. The surface at each point along the track either is smooth or has a tiny pit. These pitted or smooth spots are used to encode the numeric values of the wave; each spot is a single bit, and a sequence of bits represents the numeric value in a binary encoding, as we will discuss in the next section. As the disk rotates, a laser shines on the track and a photoelectric sensor detects changes in how much light is reflected. If there's not much light, there was a pit; if there's a lot of reflected light, there was no pit. The standard encoding for CDs uses 44,100 samples per second; each sample is two values (left side and right side, for stereo) of amplitudes measured to an accuracy of one part in 65,536, which is 2^{16}. The pits are so small that they can only be seen with a microscope. DVDs are similar; smaller spots and a shorter wavelength laser allow them to store nearly 5 GB, compared to about 700 MB for a CD.

The audio CD almost drove the LP out of existence because it was so much better in most ways—not subject to wear because there is no physical contact from the laser, not much bothered by dirt or scratches, not fragile, and definitely compact. LPs periodically enjoy a modest renaissance, while CDs of popular music are in serious decline because it's easier and cheaper to download music from the Internet. CDs had a second career as a storage and distribution medium for software and data, but that was superseded by DVDs, which in turn have largely been replaced by Internet storage and downloading. To many readers, audio CDs may seem as antique as LPs. Nevertheless I am happy that my music collection is entirely on CDs (though they are also stored in MP3 format on removable hard drives). I own them outright, which is not true of music collections "in the cloud." And manufactured CDs will outlast me, though copied ones may not, because they rely on a chemical change in a light-sensitive dye whose properties may change over time.

Because they contain more detail than humans can perceive, sound and images can be *compressed*. For music, this is done with compression techniques like MP3 and AAC (Advanced Audio Coding), which reduce the size by a factor of 10 with little perceptible degradation. For images, the most common compression technique is called JPEG, named after the Joint Photographic Experts Group, the organization that defined it; it also shrinks an image by a factor of 10 or more. Compression is an example of the kind of processing that can be done on digital information but would be extremely difficult if not impossible with analog. We'll discuss compression further in Chapter 8.

2.2.3 Digitizing movies

What about movies? In the 1870s, the English photographer Eadweard Muybridge showed how to create the illusion of motion by displaying a sequence of still images one after another in rapid succession. Today, a motion picture displays images at 24 frames per second, and TV displays at 25 to 30 frames per second, which is fast enough that the human eye perceives the sequence as continuous motion. Video games are typically 60 frames per second. Old movies used only a dozen frames per second so they had noticeable flicker; that artifact lives on in the old word "flicks" for movies and today in the name Netflix.

A digital representation of a movie combines and synchronizes the sound and picture components. Compression can be used to reduce the amount of space required, as in standard movie representations like MPEG ("Moving Picture Experts Group"). In practice, video representation is more complicated than audio, in part because it's intrinsically harder, but also because much of it is based on standards for broadcast television, which for most of its lifetime was analog. Analog television is being phased out in most parts of the world. In the US, television broadcasting switched to digital in 2009; other countries are in various stages of the process.

Movies and television shows are a combination of pictures and sound, and commercial ones cost much more to produce than music does. Yet it's just as easy to make perfect digital copies and send them around the world for free. So the copyright stakes are higher than for music, and the entertainment industry continues its campaign against copying.

2.2.4 Digitizing text

Some kinds of information are easy to represent in digital form, since no transformation is needed beyond agreement on what the representation is. Consider ordinary text, like the letters, numbers and punctuation in this book. We could assign a unique number to each different letter—A is 1, B is 2, and so on—and that would be a fine digital representation. In fact, that's exactly what is done, except that in the standard representation, A through Z are 65 through 90, a through z are 97 through 122, the digits 0 through 9 are 48 through 57, and other characters like punctuation take other values. This representation is called *ASCII*, the American Standard Code for

Information Interchange, which was standardized in 1963.

Figure 2.6 shows part of ASCII; I've omitted the first four rows, which contain tab, backspace and other non-printing characters.

32	space	33	!	34	"	35	#	36	$	37	%	38	&	39	'
40	(	41	)	42	*	43	+	44	,	45	–	46	.	47	/
48	0	49	1	50	2	51	3	52	4	53	5	54	6	55	7
56	8	57	9	58	:	59	;	60	<	61	=	62	>	63	?
64	@	65	A	66	B	67	C	68	D	69	E	70	F	71	G
72	H	73	I	74	J	75	K	76	L	77	M	78	N	79	O
80	P	81	Q	82	R	83	S	84	T	85	U	86	V	87	W
88	X	89	Y	90	Z	91	[	92	\	93	]	94	^	95	_
96	`	97	a	98	b	99	c	100	d	101	e	102	f	103	g
104	h	105	i	106	j	107	k	108	l	109	m	110	n	111	o
112	p	113	q	114	r	115	s	116	t	117	u	118	v	119	w
120	x	121	y	122	z	123	{	124			125	}	126	~	127	del

Figure 2.6: ASCII characters and their numeric values.

There are multiple character-set standards in different geographic or linguistic regions, but the world has more or less converged on a single standard called Unicode, which specifies a unique numeric value for every character in every language. This is a big collection, since humans have been endlessly inventive but rarely systematic in their creation of writing systems. Unicode has over 140,000 characters and the number rises steadily. As might be imagined, Asian character sets like Chinese account for a substantial part of Unicode, but by no means all. The Unicode web site, `unicode.org`, has charts of all the characters; it's fascinating and well worth a detour.

The bottom line: a digital representation can represent all of these kinds of information and indeed anything that can be converted into numeric values. Since it is just numbers, it can be processed by digital computers; as we will see in Chapter 9 it can be copied to any other computer by the universal digital network, the Internet.

2.3 Bits, Bytes, and Binary

"There are only 10 kinds of people in the world—those who understand binary numbers and those who don't."

Digital systems represent information of all types as numeric values, but perhaps surprisingly, they do not use the familiar base ten (decimal) number system internally. Instead they use binary numbers, that is, numbers in base two.

Although everyone is more or less comfortable with arithmetic, in my experience their understanding of what a number means is sometimes shaky, at least when it comes to drawing the analogy between base ten (totally familiar) and base two (not familiar to most). I'll try to remedy this problem in this section, but if things seem confused or confusing, keep repeating to yourself, "It's just like ordinary numbers, but with two instead of ten."

2.3.1 Bits

The most elemental way to represent digital information is with bits. As noted in the quotation at the beginning of this chapter, the word *bit* is a contraction of *binary digit* that was coined by the statistician John Tukey in the mid-1940s. It is said that Edward Teller, best known as the father of the hydrogen bomb, preferred "bigit," a term that mercifully didn't catch on.

The word "binary" suggests something with two values (the prefix "bi" means two), and that is indeed the case: a bit is a digit that takes on either the value 0 or the value 1, with no other possibilities. This can be contrasted with the 10 possible values of the decimal digits 0 through 9.

With a single bit, we can encode or represent any choice that involves selecting one of two values. Such binary choices abound: on/off, true/false, yes/no, high/low, in/out, up/down, left/right, north/south, east/west, and so on. A single bit is sufficient to identify which one of the pair was selected. For example, we could assign 0 to off and 1 to on, or vice versa, so long as everyone agrees on which value represents which state.

Figure 2.7 shows the power switch on my printer and the standard on-off symbol that is seen on many devices. It's a Unicode character too.

Figure 2.7:   On-off switch and standard on-off symbol.

A single bit is enough to represent on/off, true/false and similar binary choices but we need a way to deal with more choices or to represent more complicated things. For that, we use a group of bits, assigning meanings to the different possible combinations of 0's and 1's. For example, we could use two bits to represent the four years of college in the US—freshman (00), sophomore (01), junior (10) and senior (11). If there were one more category, say graduate student, two bits is not sufficient: there would be five possible values but only four different combinations of two bits. Three bits would be enough, however, and in fact could represent as many as eight different kinds of things, so we could also include faculty, staff and post-doc. The combinations would be 000, 001, 010, 011, 100, 101, 110, and 111.

There's a pattern that relates the number of bits to the number of items that can be labeled with that many bits. The relationship is simple: if there are N bits, the number of different bit patterns is 2^N, that is, $2 \times 2 \times \cdots \times 2$ (N times), as in Figure 2.8.

number of bits	number of values		number of bits	number of values
1	2		6	64
2	4		7	128
3	8		8	256
4	16		9	512
5	32		10	1,024

Figure 2.8: Powers of 2.

This is directly analogous to decimal digits: with N decimal digits, the number of different digit patterns (which we call "numbers") is 10^N, as in Figure 2.9.

number of digits	number of values		number of digits	number of values
1	10		6	1,000,000
2	100		7	10,000,000
3	1,000		8	100,000,000
4	10,000		9	1,000,000,000
5	100,000		10	10,000,000,000

Figure 2.9: Powers of 10.

2.3.2 Powers of two and powers of ten

Since everything in a computer is handled in binary, properties like sizes and capacities tend to be expressed in powers of two. If there are N bits, there are 2^N possible values, so it's handy to know the powers of two up to some value, say 2^{10}. Once the numbers get larger, they are certainly not worth memorizing. Fortunately, Figure 2.10 shows that there's a shortcut that gives a good approximation: certain powers of two are close to powers of ten, in an orderly way that's easy to remember. Figure 2.10 includes one more size prefix, "peta" or 10^{15}; it's pronounced like "pet," not "Pete." The glossary at the end of the book has a larger table with still more units.

$$2^{10} = 1,024 \qquad\qquad 10^3 = 1,000 \ \text{ (kilo)}$$
$$2^{20} = 1,048,576 \qquad\qquad 10^6 = 1,000,000 \ \text{ (mega)}$$
$$2^{30} = 1,073,741,824 \qquad\qquad 10^9 = 1,000,000,000 \ \text{ (giga)}$$
$$2^{40} = 1,099,511,627,776 \qquad\qquad 10^{12} = 1,000,000,000,000 \ \text{ (tera)}$$
$$2^{50} = 1,125,899,906,842,624 \qquad 10^{15} = 1,000,000,000,000,000 \ \text{ (peta)}$$
$$\dots$$

Figure 2.10: Powers of 2 and powers of 10.

The approximation gets worse as the numbers get bigger, but it's only 12.6 percent too high at 10^{15}, so it's useful over a very wide range. You'll find that people often blur the distinction between the power of two and the power of ten (sometimes in a direction that favors some point that they are trying to make), so "kilo" or "1K" might mean one thousand, or it might mean 2^{10} or 1,024. This is usually a minor difference, so the powers of two and ten are a good way to do mental arithmetic on big numbers involving bits.

2.3.3 Binary numbers

A sequence of bits can represent a numeric value if the digits are interpreted in the usual place-value sense, but using base 2 instead of base 10. Ten digits, 0 through 9, are enough to assign labels to up to ten items. If we need to go beyond ten, we must use more digits; with two decimal digits, we can label up to 100 things, with labels 00 through 99. For more than 100 items, it's on to three digits, which gives a range of 1,000, from 000 to 999. We don't normally write the leading zeroes for ordinary numbers, but they are there implicitly. In daily life we also start labeling at one, not zero.

Decimal numbers are shorthands for sums of powers of ten; for example, 1867 is $1\times10^3 + 8\times10^2 + 6\times10^1 + 7\times10^0$, which is $1\times1000 + 8\times100 + 6\times10 + 7\times1$, which is 1000+800+60+7. In elementary school, you might have called these the 1's column, the 10's column, the 100's column and so on. This is so familiar that we rarely think about it.

Binary numbers are the same except that the base is two instead of ten, and the only digits involved are 0 and 1. A binary number like 11101 is interpreted as $1\times2^4 + 1\times2^3 + 1\times2^2 + 0\times2^1 + 1\times2^0$, which we express in base ten as 16+8+4+0+1 or 29.

The fact that sequences of bits can be interpreted as numbers means that there is a natural pattern for assigning binary labels to items: put them in numeric order. We saw that above with the labels for freshmen, sophomores, and so on: 00, 01, 10, 11, which in base ten have the numeric values 0, 1, 2 and 3. The next sequence would be 000, 001, 010, 011, 100, 101, 110, 111, with numeric values from 0 to 7.

Here's an exercise to confirm your understanding. We're all familiar with counting up to ten on our fingers, but how high can you count on your fingers if you use binary numbers, with each finger and thumb (a digit!) representing a binary digit? What's the range of values? If the binary representations of 4 and 132 remind you of something, you've got the idea.

As you can see, it's easy to convert binary to decimal: just add up the powers of two for which the corresponding bit of the number is 1. Converting decimal to binary is harder, but not by much. Repeatedly divide the decimal number by two. Write down the remainder, which will be either zero or one, and use the quotient as the next value to divide. Keep going until the original number has been divided down to zero. The sequence of remainders is the binary number, except it's in reverse order, so flip it end for end.

As an example, Figure 2.11 shows the conversion of 1867 to binary. Reading the bits off backwards, we have 111 0100 1011, which we can check by adding up the powers of two: 1024+512+256+64+8+2+1 = 1867.

Each step of this procedure produces the least significant (rightmost) bit of the number that remains. It's analogous to the process you would go through to convert a large number in seconds into days, hours, minutes and seconds: divide by 60 to get minutes (and the remainder is the number of seconds); divide the result by 60 to get hours (and the remainder is the number of minutes); divide the result by 24 to get

number	quotient	remainder
1867	933	**1**
933	466	**1**
466	233	**0**
233	116	**1**
116	58	**0**
58	29	**0**
29	14	**1**
14	7	**0**
7	3	**1**
3	1	**1**
1	0	**1**

Figure 2.11: Conversion of decimal 1867 to binary 11101001011.

days (and the remainder is the number of hours). The difference is that time conversions do not use a single base, but mix bases of 60 and 24.

You can also convert decimal to binary by subtracting decreasing powers of two from the original number, starting with the highest power of two that the number contains, like 2^{10} in 1867. Each time a power is subtracted, write 1, or if the power is too big, write 0, as with 2^7 or 128 in the example above. At the end the sequence of 1's and 0's will be the binary value. This approach is perhaps more intuitive but not so mechanical.

Binary arithmetic is easy. Since there are only two digits to work with, the addition and multiplication tables have only two rows and two columns, as shown in Figure 2.12. It's unlikely that you'll ever need to do binary arithmetic yourself, but the simplicity of these tables hints at why computer circuitry for binary arithmetic is much simpler than it would be for decimal arithmetic.

+	0	1		×	0	1
0	0	1		0	0	0
1	1	0 and carry 1		1	0	1

Figure 2.12: Binary addition and multiplication tables.

2.3.4 Bytes

In all modern computers, the basic unit of processing and memory organization is 8 bits that are treated as a unit. A group of eight bits is called a *byte*, a word coined in 1956 by Werner Buchholz, a computer architect at IBM. A single byte can encode 256 distinct values (2^8, all the different combinations of 8 zeroes and ones), which could be an integer value between 0 and 255, or a single character in the 7-bit ASCII character set (with one bit to spare), or something else. Often, a particular byte is part of a larger group that represents something bigger or more complicated. Two bytes together provides 16 bits, enough to represent a number between 0 and $2^{16} - 1$, or 65,535. They could also represent a character in the Unicode character set, perhaps one or the other of

東京

which are the two Unicode characters of "Tokyo"; each character is two bytes. Four bytes is 32 bits, which could be four ASCII characters, or two Unicode characters, or a number up to $2^{32} - 1$, around 4.3 billion. There's no limit to what a set of bytes might represent, though the processor itself has a modest set of specific groupings, like integers of various sizes, and has instructions for processing such groups.

If we want to write down the numeric value represented by one or more bytes, we could express it in decimal form, which is convenient for a human reader if it's really numeric. We could write it in binary to see the individual bits, which matters if different bits encode different kinds of information. Binary is bulky, however, more than three times longer than the decimal form, so an alternative notation called *hexadecimal* is commonly used. Hexadecimal is base 16, so it has 16 digits (like decimal has 10 digits and binary has 2); the digits are 0, 1, ..., 9, A, B, C, D, E, and F. Each hex digit represents 4 bits, with the numeric values shown in Figure 2.13.

```
0   0000   1   0001   2   0010   3   0011
4   0100   5   0101   6   0110   7   0111
8   1000   9   1001   A   1010   B   1011
C   1100   D   1101   E   1110   F   1111
```

Figure 2.13: Table of hexadecimal digits and their binary values.

Unless you're a programmer, there are only a handful of places where you might see hexadecimal. One is colors on a web page. As mentioned earlier, the most common representation of colors in a computer uses three bytes for each pixel, one for the amount of red, one for the amount of green, and one for the amount of blue; this is called RGB encoding. Each of those components is stored in a single byte, so there are 256 possible amounts of red; for each of those there are 256 possible amounts of green, and then for each of those there are 256 possible blue values. Altogether that's $256 \times 256 \times 256$ possible colors, which sounds like a lot. We can use powers of two and ten to get a quick estimate of how many. It's $2^8 \times 2^8 \times 2^8$, which is 2^{24}, or $2^4 \times 2^{20}$, or about 16×10^6, or 16 million. You've probably seen this number used to describe computer displays ("More than 16 million colors!"). The estimate is about 5 percent low; the true value of 2^{24} is 16,777,216.

An intense red pixel would be represented as FF0000, that is, maximum red of 255 decimal, no green, no blue, while a bright but not intense blue, like the color of links on many web pages, would be 0000CC. Yellow is red plus green, so FFFF00 would be the brightest possible yellow. Shades of gray have equal amounts of red, green and blue, so a medium gray pixel would be 808080, that is, the same amount of red, green and blue. Black and white are 000000 and FFFFFF respectively.

Hex values are also used in Unicode code tables to identify characters:

東京

is 6771 4EAC. You will also see hex in Ethernet addresses, which we'll talk about in Chapter 8, and representing special characters in URLs, in Chapter 10.

You'll sometimes see the phrase "64-bit" in computer advertisements ("Microsoft Windows 10 Home 64-bit"). What does this mean? Computers manipulate data internally in chunks of different sizes; the chunks include numbers, for which 32 bits and 64 bits are convenient, and addresses, that is, locations of information in primary memory. It is this latter property that is being referred to. Thirty years ago there was a transition from 16-bit addresses to 32-bit addresses, which are big enough to access up to 4 GB of memory, and today the transition from 32 to 64 bits for general-purpose computers is almost complete. I won't try to predict when the transition from 64 to 128 will occur, but we should be safe for a while.

The critical thing to remember in all of this discussion of bits and bytes is that the meaning of a group of bits depends on their context; you can't tell what they mean just by looking at them. A single byte could be a single bit representing true or false and 7 unused bits, or it could be storing a small integer or an ASCII character like #, or it could be part of a character in another writing system, part of a larger number with 2 or 4 or 8 bytes, or part of a picture or a piece of music, or part of an instruction for the processor to execute, or many other possibilities. (This is just like decimal digits. Depending on context, a three-decimal-digit number could represent a US area code, a highway number, a baseball batting average, or many other things.)

One program's instructions are sometimes another program's data. When you download a program or app, it's just data: bits to be copied blindly. But when you run the program, its bits are treated as instructions as they are processed by the CPU.

2.4 Summary

Why binary instead of decimal? The answer is that it's much easier to make physical devices that have only two states, like on and off, than devices that have ten states. This comparative simplicity is exploited in many technologies: current (flowing or not), voltage (high or low), electric charge (present or not), magnetism (north or south), light (bright or dim), reflectance (shiny or dull). Von Neumann clearly realized this; in 1946 he said, "Our fundamental unit of memory is naturally adapted to the binary system since we do not attempt to measure gradations of charge."

Why should anyone know or care about binary numbers? One reason is that working with numbers in an unfamiliar base is an example of quantitative reasoning that might even improve understanding of how numbers work in good old base ten. Beyond that, it's also important because the number of bits is usually related in some way to how much space, time or complexity is involved. And fundamentally, computers are worth understanding, and binary is central to their operation.

Binary shows up in real-world settings unrelated to computing as well, probably because doubling or halving weights, lengths, and so on is such a natural operation for people. For instance, volume 2 of Donald Knuth's *The Art of Computer Programming* describes English wine container units in the 1300s that run over 13 binary orders of magnitude: 2 gills is one chopin, 2 chopins is one pint, 2 pints is one quart, and so on until 2 barrels is one hogshead, 2 hogsheads is one pipe, and 2 pipes is one tun. About half of those units are still in common use in the English system of liquid measures, though charming words like firkin and kilderkin (two firkins, or half a barrel) are now rarely seen.

3

Inside the Processor

"If, however, the orders to the machine are reduced to a numerical code and if the machine can in some fashion distinguish a number from an order, the memory organ can be used to store both numbers and orders."

Arthur W. Burks, Herman H. Goldstine, John von Neumann, "Preliminary discussion of the logical design of an electronic computing instrument," 1946.

In Chapter 1, I said that the processor or CPU was "the brain" of a computer, though with a caveat that the term didn't really make sense. It's now time to take a detailed look at the processor, since it is the most important component of a computer, the one whose properties are the most significant for the rest of the book.

How does the processor work? What does it process and how? To a first approximation, the processor has a repertoire of basic operations that it can perform. It can do arithmetic—add, subtract, multiply and divide numbers, like a calculator. It can fetch data from the memory to operate on and it can store results back into the memory, like the memory operations on many calculators. And it controls the rest of the computer; it uses signals on the bus to orchestrate and coordinate input and output for whatever is electrically connected to it, including mouse, keyboard, display, and everything else.

Most important, it can make decisions, albeit of a simple kind: it can compare numbers (is this number bigger than that one?) or other kinds of data (is this piece of information the same as that one?), and it can decide what to do next based on the outcome. This is the most important thing of all, because it means that although the processor can't do much more than a calculator can, it can operate without human intervention. As Burks, Goldstine and von Neumann said, "It is intended that the machine be fully automatic in character, i.e., independent of the human operator after the computation starts."

Because the processor can decide what to do next based on the data that it is processing, it is able to run the whole system on its own. Although its repertoire is not large or complicated, the processor can perform billions of operations every second, so it can do exceedingly sophisticated computations.

3.1 The Toy Computer

Let me explain how a processor operates by describing a machine that doesn't exist. It's a made-up or "pretend" computer that uses the same ideas as a real computer but is much simpler. Since it exists only on paper, I can design it in any way that might help to explain how real computers work. I can also create a program for a real computer that will *simulate* my paper design, so I can write programs for the pretend machine and see how they run.

I've taken to calling this made-up machine the "Toy" computer, since it's not real but it has many of the properties of the real thing; it's at about the level of minicomputers of the late 1960s, and somewhat similar to the design presented in the original paper by Burks, Goldstine and von Neumann. The Toy has memory for storing instructions and data, and it has one additional storage area called the *accumulator* that has enough capacity to hold a single number. The accumulator is analogous to the display in a calculator, which holds the number most recently entered by the user or the most recent computational result. The Toy has a repertoire of about ten instructions for performing basic operations like those described above. Figure 3.1 shows the first six.

GET	get a number from the keyboard into the accumulator, overwriting the previous contents
PRINT	print the contents of the accumulator (accumulator contents do not change)
STORE M	store a copy of the accumulator contents into memory location M (accumulator contents do not change)
LOAD M	load the accumulator with contents of memory location M (contents of M do not change)
ADD M	add contents of memory location M to contents of accumulator (contents of M do not change)
STOP	stop execution

Figure 3.1: Representative Toy machine instructions.

Each memory location holds one number or one instruction, so a program consists of a sequence of instructions and data items stored in the memory. In operation, the processor starts at the first memory location and repeats a simple cycle:

Fetch:	get the next instruction from memory
Decode:	figure out what that instruction does
Execute:	perform the instruction go back to *Fetch*

3.1.1 The first Toy program

To create a program for the Toy, we have to write a sequence of instructions that will do the desired task, put them into the memory, and tell the processor to start executing them. As an example, suppose that the memory contains exactly these instructions, which would be stored in memory as binary numbers:

```
GET
PRINT
STOP
```

When this sequence is executed, the first instruction will ask the user for a number, the second will print that number, and the third will tell the processor to stop. This is terminally boring, but it is enough to show what a program looks like. Given a real Toy machine, the program could even be run.

Fortunately, there are working Toy computers. Figure 3.2 shows one of them in operation; it's a simulator that was written in JavaScript so that it can run in any browser, as we will see in Chapter 7.

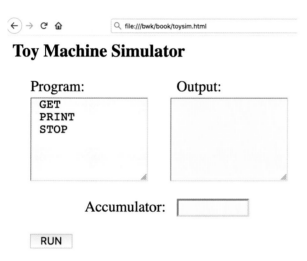

Figure 3.2: Toy machine simulator with a program ready to run.

When you push RUN, the dialog box of Figure 3.3 pops up when the GET instruction is executed; the number 123 has been typed by the user.

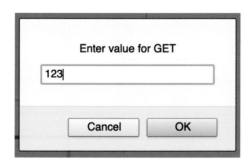

Figure 3.3: Toy machine simulator input dialog box.

After the user types a number and pushes OK, the simulator runs and displays the result shown in Figure 3.4. As promised, the program asks for an input number, prints it, and stops.

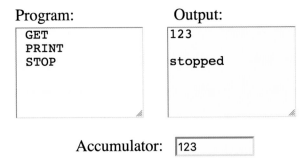

Figure 3.4: Toy machine simulator after running a short program.

3.1.2 The second Toy program

The next program (Figure 3.5) is slightly more complicated and adds a new idea: storing a value in the memory and then retrieving it. The program gets a number into the accumulator, stores it in memory, gets a second number into the accumulator (which overwrites the first number), adds the first number to it (taken from memory, where we had carefully stored it), prints the sum of the two numbers, and then stops.

The processor starts at the beginning of the program and fetches instructions one at a time. It executes each instruction in turn, then goes on to the next one. Each instruction is followed by a *comment*, that is, explanatory material to help programmers; comments have no effect on the program itself.

`GET`	get first number into the accumulator
`STORE FirstNum`	store it in a memory location called `FirstNum`
`GET`	get second number into the accumulator
`ADD FirstNum`	add first number to value in the accumulator
`PRINT`	print resulting sum
`STOP`	stop running the program
`FirstNum:`	a location in memory to hold the first input number

Figure 3.5: Toy machine program to add two numbers and print the sum.

The only tricky bit is that we need to set aside a place in the memory to hold a data value, the first number that will be read. We can't leave the first number in the accumulator, since the second `GET` instruction will overwrite it. Because the number is data, not an instruction, we have to store it someplace in the memory where it won't be interpreted as an instruction. If we place it at the end of the program, after all of the instructions, the processor will never try to interpret the data value as an instruction because it will `STOP` before it gets there.

We also need a way to refer to that location when it's needed by instructions of the program. One possibility would be to observe that the data will be in the seventh memory location (after the six instructions), so we could write "`STORE 7`". In fact the program will eventually be stored in this form. But the location might change if the program is modified. The solution is to give the data location a name, and as we'll see in Chapter 5, a *program* can do the clerical task of keeping track of where the data is located in the memory, replacing the name by the proper numeric location. The name FirstNum is meant to suggest that it's the "first number." The name is arbitrary, though it's good practice to use a name that indicates the purpose or meaning of the data or instruction it is attached to. We use a colon after the name to indicate that it is a label. By convention, instructions in a program are indented while the names attached to instructions or memory locations are not indented. The Toy simulator takes care of all these details.

3.1.3 Branch instructions

How could the program of Figure 3.5 be extended so it will add three numbers? It would be easy enough to add another sequence of STORE, GET and ADD instructions (there are two places they could be inserted), but that certainly won't scale up to adding a thousand numbers, nor would it work if we don't know in advance how many numbers there will be.

The solution is to add a new kind of instruction to the processor's repertoire that lets it re-use sequences of instructions. The GOTO instruction, often called "branch" or "jump," tells the processor to take its next instruction not from the next one in sequence but from the location specified in the GOTO instruction itself.

With a GOTO instruction we can make the processor return to an earlier part of the program and repeat instructions. One simple example is a program that prints each number as it is entered; that's the essence of a program that copies or displays its input, and it shows what the GOTO instruction does. The first instruction of the program in Figure 3.6 is labeled Top, an arbitrary name that suggests its role, and the last instruction causes the processor to go back to that first instruction.

```
Top:  GET         get a number into the accumulator
      PRINT       print it
      GOTO Top    go back to Top to get another number
```

Figure 3.6: Data-copying program that runs forever.

This gets us part of the way—we can re-use instructions—but there's still a critical problem: there's no way to stop this repeated sequence of instructions, or *loop*, from continuing indefinitely. To stop the loop we need another kind of instruction, one that can test a condition and decide what to do next rather than pressing on blindly. That kind of instruction is called a *conditional branch* or *conditional jump*. One possibility offered by all computers is an instruction that tests whether a value is zero, and jumps to a specific instruction if it is. Fortunately, the Toy has an instruction called IFZERO that branches to a specified instruction if the accumulator value is zero; otherwise, execution continues at the next sequential instruction.

We can use the IFZERO instruction to write a program (Figure 3.7) that reads and prints input values until a value of zero appears in the input.

This keeps fetching data and printing it until the user gets tired and enters a zero, at which point the program jumps to the STOP instruction, labeled Bot for "bottom," and quits. (It's tempting to write IFZERO STOP but that won't work: IFZERO must be followed by a location, not an instruction.)

```
Top: GET            get a number into the accumulator
     IFZERO Bot     if the value in the accumulator is zero, go to instruction Bot
     PRINT          accumulator value was not zero, so print it
     GOTO Top       go back to Top to get another number
Bot: STOP
```

Figure 3.7: Data-copying program that stops when 0 is entered.

Note that the program does not print the zero that signaled the end of the input. How would you modify the program so that it does print the zero before stopping? This is not a trick question—the answer is clear—but it is a good illustration of how programs may embody subtle distinctions in what they are supposed to do or how they might do something different than what was intended, all because of a simple transposition of two instructions.

The combination of GOTO and IFZERO lets us write programs that repeat instructions until some specified condition becomes true; the processor can change the course of a computation according to the results of previous computations. (You might think about whether GOTO is strictly necessary if you have IFZERO—is there a way to simulate GOTO with an IFZERO and some other instructions?) It's not obvious, but this is all we need to compute anything that can be computed by any digital computer—any computation can be broken down into small steps using elementary instructions.

With IFZERO in its repertoire, the Toy processor can in principle be programmed to do literally *any* computation.

I say "in principle" because in practice we can't ignore processor speed, memory capacity, the finite sizes of numbers in a computer, and the like. We'll come back to this idea of the equivalence of all computers from time to time, since it's a fundamental notion.

As another example of IFZERO and GOTO, Figure 3.8 shows a program that will add up a bunch of numbers, stopping when the number zero is entered. Using a special value to terminate a sequence of inputs is a common practice. Zero works well as an end marker in this specific example because we're adding up numbers and adding a zero data value isn't necessary.

```
Top: GET            get a number
     IFZERO Bot     if it's zero, go to Bot
     ADD Sum        add running sum to the latest number
     STORE Sum      store result as new running sum
     GOTO Top       go back to Top to get another number
Bot: LOAD Sum       load running sum into accumulator
     PRINT          and print it
     STOP
Sum: 0              memory location to hold running sum
                      (initialized to 0 when program starts)
```

Figure 3.8: Toy machine program to add up a sequence of numbers.

Toy simulators interpret an "instruction" like the last line of this program to mean "assign a name to a memory location and put a specific numeric value in that location before the program starts to run." It's not a real instruction, but rather a "pseudo-instruction" that is interpreted by the simulator as it is processing the program text, before it starts to run the program.

We need a place in memory to hold the running sum as it is being added to. That memory location should start out holding the value zero, just like clearing the memory in a calculator. We also need a name for the memory location that can be used by the rest of the program to refer to it. That name is arbitrary but Sum is a good choice since it indicates the role of the memory location.

How would you check this program to be sure it works? It looks OK on the surface and it produces the right answers on simple test cases, but problems are easily overlooked, so it's important to test systematically. The operative word is "systematically"; it's not effective to just throw random inputs at a program.

What's the simplest test case? If there are no numbers at all except the zero that terminates the input, the sum should be zero, so that's a good first test case. The second thing to try is to input a single number; the sum should be that number. Next is to try two numbers whose sum you already know, like 1 and 2; the result should be 3. With a few tests like this, you can be reasonably confident that the program is working. If you're careful, you can test the code before it ever gets near a computer, by stepping carefully through the instructions yourself. Good programmers do this kind of checking for every program they write.

3.1.4 Representation in memory

So far I have ducked the question of exactly how instructions and data are represented in the memory. How might that work?

Here's one possibility. Suppose that each instruction uses one memory location to store its numeric code and also uses the next location if the instruction refers to memory or has a data value. That is, GET occupies one location while instructions like IFZERO and ADD that refer to a memory location occupy two memory cells, the second of which is the location that they refer to.

Suppose also that any data value will fit in a single location. These are simplifications, though not far off from what happens in real computers. Finally, suppose that the numeric values of the instructions are GET = 1, PRINT = 2, STORE = 3, LOAD = 4, ADD = 5, STOP = 6, GOTO = 7, IFZERO = 8, following their order of appearance in previous pages.

The program in Figure 3.8 adds up a sequence of numbers. When the program is about to begin, the memory contents would be as shown in Figure 3.9, which also shows the actual memory locations, the labels attached to three locations, and the instructions and addresses that correspond to the memory contents.

The Toy simulator is written in JavaScript, a programming language that we will talk about in Chapter 7, though it could be written in any language. The simulator is easy to extend. For instance, it's straightforward to add a multiplication instruction or a different kind of conditional branch even if you've never seen a computer program before; it's a good way to test your understanding. The code can be found on the web site for the book.

Location	Memory	Label	Instruction
1	1	Top:	GET
2	8		IFZERO Bot
3	10		
4	5		ADD Sum
5	14		
6	3		STORE Sum
7	14		
8	7		GOTO Top
9	1		
10	4	Bot:	LOAD Sum
11	14		
12	2		PRINT
13	6		STOP
14	0	Sum:	0 [data, initialized to 0]

Figure 3.9: Add-up-the-numbers program in memory.

3.2 Real Processors

What we've just seen is a simplified version of a processor, though not too unrealistic for early or small computers. Reality today is much more complex in the details, centering around performance.

A processor performs the fetch, decode, execute cycle over and over. It fetches the next instruction from memory, which is normally the instruction stored in the next memory location, but could instead be one from a location specified by a GOTO or IFZERO. The processor decodes the instruction, that is, it figures out what the instruction does and makes whatever preparations are necessary to carry it out. It then executes the instruction, by fetching information from memory, doing arithmetic or logic, and storing results, in whatever combination is appropriate for that instruction. Then it goes back to the fetch part of the cycle. The fetch-decode-execute cycle in a real processor has elaborate mechanisms for making the whole process run fast, but at its heart it's just a loop like the ones we showed above for adding up numbers.

Real computers have more instructions than our Toy does, though the instructions are of the same basic types. They have more ways to move data around, more ways to do arithmetic and on different sizes and kinds of numbers, more ways to compare and branch, and more ways to control the rest of the machine. A typical processor will have a few dozen to a few hundred different instructions; instructions and data occupy several memory locations, often 2 to 8 bytes. A real processor will have multiple accumulators, often 16 or 32, so that it can hold more than one intermediate result in what amounts to extremely fast memory.

Real programs are enormous by comparison with our Toy examples, often with millions of instructions. We'll come back to how such programs are written when we talk about software in later chapters.

Computer architecture is the discipline that deals with designing the processor and its connections to the rest of the machine; in universities, it's often a subfield on the border of computer science and electrical engineering.

One concern of computer architecture is the instruction set—the repertoire of instructions that the processor provides. Should there be a large number of instructions to handle a wide variety of different kinds of computation, or should there be fewer instructions that would be simpler to build and might run faster? Architecture involves complicated tradeoffs among functionality, speed, complexity, power consumption, and programmability. Von Neumann again: "In general, the inner economy of the arithmetic unit is determined by a compromise between the desire for speed of operation [...] and the desire for simplicity, or cheapness, of the machine."

How is the processor connected to the primary memory and the rest of the computer? Processors are very fast, performing an instruction in well under a nanosecond. (Recall that "nano" is one billionth, or 10^{-9}.) By comparison, memory is excruciatingly slow—fetching data or an instruction from memory might take 10 to 20 nanoseconds. That's fast in absolute terms, of course, but it's slow from the perspective of the processor, which might have executed dozens of instructions if it didn't have to wait for data to arrive.

Modern computers use a small number of high-speed memories called *caches* between the processor and the memory to hold recently used instructions and data. Accessing information that can be found in a cache is faster than waiting for it to come from primary memory. I'll discuss caches and caching in the next section.

Designers also have a collection of architectural tricks that make processors run faster. A processor can be designed to overlap fetch and execute so that there are several instructions in various stages of completion; this is called pipelining, and is similar in spirit to cars moving along an assembly line. The result is that although any given instruction still takes the same time to complete, others are in process at the same time and the overall completion rate is higher. Another option is to do multiple instructions in parallel if they do not interfere with or depend on each other; the automotive analogy would be to have parallel assembly lines. Sometimes it's even possible to do instructions out of order if they do not interact.

Yet another option is to have multiple processors operating at the same time. This is the norm in laptops and cell phones today. The Intel processor in the 2015 computer I'm using right now has two cores on a single integrated circuit chip, and the trend is strongly towards more and more processor cores on single chips and more than one chip per machine. As integrated-circuit feature sizes become smaller, it's possible to pack more transistors onto a chip, and those tend to be used for more cores and more cache memory. Individual processors are not getting faster but there are more cores so effective computation speeds are still rising.

A different set of tradeoffs in processor design comes into play when one considers where the processor will be used. For a long time, the main target was desktop computers, where electrical power and physical space were comparatively plentiful. That meant that designers could concentrate on making the processor run as fast as possible, knowing that there would be plenty of power and a way to dissipate heat with fans. Laptops significantly changed this set of tradeoffs, since space is tight and an unplugged laptop gets its power from a battery that is heavy and expensive. All else being equal, processors for laptops tend to be slower and use less power.

Cell phones, tablets and other highly portable devices push this tradeoff even further, since size, weight and power are even more constrained. This is an area where merely tweaking a design is not good enough. Although Intel and its primary competitor AMD are the dominant suppliers of processors for desktops and laptops, most cell phones and tablets use a processor design called "ARM," which is specifically designed to use low power. ARM processor designs are licensed from the English company Arm Holdings.

Processor speed comparisons are difficult and not terribly meaningful. Even basic operations like arithmetic can be handled in ways that are sufficiently different that it's hard to make a head-to-head comparison. For example, one processor might require three instructions to add two numbers and store the result in a third, as the Toy does. A second processor might require two instructions, while a third processor might have a single instruction for that computation. One processor might be able to handle several instructions in parallel or overlap several so that they are proceeding in stages. Processors can sacrifice speedy execution for lower power consumption, even adjusting their speed dynamically depending on whether power is coming from batteries. Some processors have some fast cores and some slow ones, allocated to different tasks. You should be cautious about claims that one processor is "faster" than another; your mileage may vary.

3.3 Caching

"We are therefore forced to recognize the possibility of constructing a hierarchy of memories, each of which has greater capacity than the preceding but which is less quickly accessible."

Arthur W. Burks, Herman H. Goldstine, John von Neumann, "Preliminary discussion of the logical design of an electronic computing instrument," 1946.

It's worth a brief digression here on caching, an idea of broad applicability far beyond computing. In the processor, a *cache* is a small, fast memory that is used to store recently used information to avoid accessing the primary memory, which is larger but much slower. A processor will normally access groups of data and instructions multiple times in short succession. For instance, the five instructions of the loop in the program of Figure 3.9 will be executed once for each input number. If those instructions are stored in a cache, then there will be no need to fetch them from memory each time through the loop and the program will run faster because it does not have to wait for the memory to produce the instructions. Similarly, keeping Sum in a data cache will speed access as well, though the real bottleneck in this program is in getting the data.

A typical processor has two or three caches, successively larger but slower, and often called levels L1, L2 and L3; the largest might hold some megabytes of data. (My laptop has 256 KB of L2 cache for each core and 4 MB in a single L3 cache.) Caching works because recently used information is likely to be used again soon—having it in the cache means less time waiting for memory. The caching process usually loads blocks of information all at once, for instance a block of consecutive memory locations when a single byte is requested. This is because adjacent information will probably be used soon as well, and thus is likely to already be in the cache when needed; references to nearby information may not have to wait.

This kind of caching is mostly invisible to users except insofar as it improves performance. But caching is a much more general idea that is helpful whenever we use something now and are likely to use it again soon or are likely to use something nearby. Multiple accumulators in processors are in effect a form of cache at the high speed end. Primary memory can be a cache for disks, and memory and disks are both caches for data coming from a network. Networks often have caches to speed up the flow of information from faraway servers, and servers themselves have caches.

You may have seen the word in the context of "clearing the cache" for a web browser. The browser keeps local copies of images and other comparatively bulky materials that are part of some web page, since it will be faster to use the local copy than to download it again if a page is revisited. The cache can't grow indefinitely, so the browser will quietly remove old items to make room for newer ones, and it offers you a way to remove everything as well.

You can sometimes observe cache effects for yourself. For example, start a big program like Word or Firefox and time how long before it finishes loading from the disk and is ready to use. Then quit the program and restart it immediately. Usually the startup will be noticeably faster the second time, because the instructions of the program are still in memory, which is serving as a cache for the disk. As you use other programs over a period of time, the memory will fill with their instructions and data. The original program won't be cached any more.

The list of recently used files in programs like Word or Excel is also a form of caching. Word remembers the files that you have used most recently and displays the names on a menu so you don't have to search to find them. As you open more files, the names of ones that have not been accessed for a while will be replaced by the more recent ones.

3.4 Other Kinds of Computers

It's easy to think that all computers are laptops, because that's what we see most often. But there are other kinds of computers, large and small, that share the core attributes of what they can logically compute, and that have similar architectures but make different tradeoffs among cost, power, size, speed, and so on.

Cell phones and tablets are computers too, running an operating system and providing a rich computing environment. Even smaller systems are embedded in almost all the digital devices that clutter our lives, including cameras, ebook readers, fitness trackers, appliances, game consoles, and on and on and on. The so-called "Internet of Things"—networked thermostats, security cameras, smart lights, voice recognizers, and the like—also relies on such processors.

Supercomputers tend to have a large number of processors and a lot of memory, and the processors themselves may have instructions that process certain kinds of data much faster than their more conventional siblings. Today's supercomputers are based on clusters of speedy but basically ordinary processors rather than specialized hardware. Every six months the web site top500.org publishes a new list of the world's 500 fastest computers. It's striking how quickly the top speed rises; a machine that might have been in the top handful a few years ago would not be on the current list at all. The top machine in November 2020, built by Fujitsu in Japan, has 7.6 million cores and can execute 537×10^{15} arithmetic operations per second at peak. Supercomputer speeds are measured by the number of *floating point operations* or *flops*, that is, arithmetic operations on numbers with a fractional part, that they can perform per second. The top of the top500.org list is thus 537 petaflops, and the 500th is 2.4 petaflops.

A *Graphics Processing Unit* or *GPU* is a specialized processor that performs certain graphics computations much faster than a general-purpose CPU. GPUs were originally developed for the high-speed graphics needed for games and are also used for speech and signal processing in phones. GPUs can also help accelerate regular processors for certain kinds of workloads. A GPU can do a large number of simple arithmetic computations in parallel, so if some part of a computational task involves operations that can be done in parallel and can be handed off to a GPU, the overall computation can proceed more rapidly. GPUs are particularly useful for machine learning (Chapter 12), where the same computation is done independently on different parts of a large dataset.

Distributed computing refers to computers that are more independent—they don't share memory, for example, and they may be physically more spread out, even located in different parts of the world. This makes communication even more of a potential bottleneck, but it enables people and computers to cooperate at large distances. Large-scale web services—search engines, online shopping, social networks, and cloud computing in general—are distributed computing systems, with thousands of computers cooperating to provide results quickly for large numbers of users.

All of these kinds of computers share the same fundamental principles. They are based on a general-purpose processor that can be programmed to perform an endless variety of tasks. Each processor has a limited repertoire of simple instructions that do arithmetic, compare data values, and select the next instruction to perform based on the results of previous computations. The general architecture hasn't changed much since the late 1940s, but the physical construction has continued to evolve at an amazing rate.

Perhaps unexpectedly, all of these computers have the same logical capabilities and can compute exactly the same things, leaving aside practical considerations like speed and memory requirements. This result was proven independently in the 1930s by several people, including the English mathematician Alan Turing. Turing's approach is the easiest for a non-specialist to understand. He described a simple computer, much simpler than our Toy, and showed that it could compute anything that was computable in a very general sense. Today that kind of computer is called a *Turing machine*. He then showed how to create a Turing machine that could simulate any other Turing machine; that's now called a *universal Turing machine*. It's easy to write a program that will simulate a universal Turing machine, and it's also possible (though not easy) to write a program for a universal Turing machine that will simulate a real computer. Hence, all computers are equivalent in what they can compute, though not in how fast they operate.

During World War II, Turing turned from theory to practice: he was central to the development of specialized computers for decrypting German military communications, which we will mention again briefly in Chapter 13. Turing's wartime work has been featured in several movies, with considerable artistic license, including *Breaking the Code* in 1996 and *The Imitation Game* in 2014.

In 1950, Turing published a paper called "Computing Machinery and Intelligence," which proposed a test (today called the *Turing test*) that one might use to assess whether a computer was displaying human intelligence. Imagine a computer and a human communicating separately with a human interrogator via a keyboard and screen. By having a conversation, can the interrogator determine which is the human and which is the computer? Turing's thought was that if they could not be reliably distinguished, the computer was displaying intelligent behavior. As we will see in Chapter 12, computers now perform at human level or above in some areas, though certainly not in anything like overall intelligence.

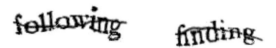

Figure 3.10: A CAPTCHA.

Turing's name is part of the somewhat forced acronym *CAPTCHA*, which stands for "Completely Automated Public Turing test to tell Computers and Humans Apart." CAPTCHAs are the distorted patterns of letters like the one in Figure 3.10 that are widely used to try to ensure that the user of a web site is a human, not a program. A CAPTCHA is an example of a *reverse Turing test*, since it attempts to distinguish between human and computer by using the fact that people are generally better at identifying the visual patterns than computers are. Of course CAPTCHAs are impossible for anyone with a visual impairment.

Turing is one of the most important figures in computing, a major contributor to our understanding of computation. The Turing Award, computer science's equivalent of the Nobel Prize, is named in his honor; later chapters will describe half a dozen important computing inventions whose inventors have received the Turing Award.

Tragically, in 1952 Turing was prosecuted for homosexual activities, which were illegal in England at the time, and he died in 1954, apparently a suicide.

3.5 Summary

A computer is a general-purpose machine. It takes its instructions from memory, and one can change the computation it performs by putting different instructions in the memory. Instructions and data are indistinguishable except by context; one person's instructions are another person's data.

A modern computer almost surely has multiple cores on a single chip and it may have several processor chips as well, with lots of cache on the integrated circuit to make memory access more efficient. Caching itself is a fundamental idea in computing, seen at all levels from the processor up through how the Internet is organized. It always involves using locality in time or space to gain faster access most of the time.

There are many ways to define the instruction set architecture of a machine, a complex tradeoff among factors like speed, power consumption, and complexity of the instructions themselves. These details are of crucial importance to hardware designers, but much less so to most of those who program the computers, and not at all to those who merely use them in some device.

Turing showed that all computers of this structure, which includes anything you're likely to see, have the same computational capabilities, in the sense that they can compute exactly the same things. Their performance can vary widely, of course, but all are equivalently capable except for issues of speed and memory capacity. The tiniest and simplest computer can in principle compute anything that its larger siblings could. Indeed, any computer can be programmed to simulate any other, which is in effect how Turing proved his result.

"It is unnecessary to design various new machines to do various computing processes. They can all be done with one digital computer, suitably programmed for each case."

Alan Turing, "Computing Machinery and Intelligence," *Mind*, 1950.

Wrap-up on Hardware

We've come to the end of our discussion of hardware, though we'll return occasionally to talk about some gadget or device. Here are the fundamental ideas you should take away from this part.

A digital computer, whether desktop, laptop, phone, tablet, ebook reader, or any of many other devices, contains one or more processors and various kinds of memory. Processors execute simple instructions very quickly. They can decide what to do next based on the results of earlier computations and inputs from the outside world. Memory contains both data and instructions that determine how to process the data.

The logical structure of computers has not changed much since the 1940s, but the physical construction has changed enormously. Moore's Law, in effect for over fifty years and by now almost a self-fulfilling prophecy, has described an exponential decrease in the size and price of individual components and thus an exponential increase in computational power for a given amount of space and money. Warnings about how Moore's Law will end in about the next ten years have been a staple of technology predictions for decades. It is clear that current integrated circuit technology is encountering trouble as devices get down to only a handful of individual atoms, but people have been remarkably inventive in the past; perhaps some new invention will keep us on the curve.

Digital devices work in binary; at the bottom level, information is represented in two-state devices because those are easiest to build and the most reliable in operation. Information of any sort is represented as collections of bits. Numbers of various types (integers, fractions, scientific notation) are represented as 1, 2, 4, or 8 bytes, sizes which computers handle naturally in hardware. That means that in ordinary circumstances numbers have a finite size and limited precision. With suitable software, it's possible to support arbitrary size and precision, though programs using such software will run more slowly. Information like characters in natural languages is also represented as some number of bytes. ASCII, which works fine for English, uses one byte per character. Less parochially, Unicode, which has several encodings, handles all character sets but uses somewhat more space. The UTF-8 encoding is a variable-length encoding of Unicode that is meant for exchanging information between systems; it uses one byte for ASCII characters and two or more bytes for others.

Analog information like measurements is converted into digital form and then back again. Music, pictures, movies, and similar types of information are converted to digital form by a sequence of measurements specific to the particular form, and converted back again for human use; in this case, some loss of information is not only expected but can be taken advantage of for compression.

Reading about hardware, and how all it does is arithmetic, might make you wonder: if the processor is no more than a high-speed programmable calculator, how can that hardware understand speech, recommend a movie you might like, or tag a friend in a photo? That's a good question. The basic answer is that even complicated processes can be broken down into tiny computational steps. We'll talk more about this in the next few chapters on software, and later on as well.

There's one last topic that should be mentioned. These are *digital* computers: everything is ultimately reduced to bits, which, individually or in groups, represent information of any sort as numbers. The interpretation of the bits depends on the context. Anything that we are able to reduce to bits can be represented and processed by a digital computer. But keep in mind that there are many, many things that we do not know how to encode in bits nor how to process in a computer. Most of these are important things in life: creativity, truth, beauty, love, honor, and values. I suspect those will remain beyond our computers for some while. You should be skeptical of anyone who claims to know how to deal with such matters "by computer."

Part II

Software

The good news is that a computer is a general-purpose machine, capable of performing any computation. Although it only has a few kinds of instructions to work with, it can do them very fast, and it can largely control its own operation.

The bad news is that it doesn't do anything itself unless someone tells it what to do, in excruciating detail. A computer is the ultimate sorcerer's apprentice, able to follow instructions tirelessly and without error, but requiring painstaking accuracy in the specification of what to do.

Software is the general term for sequences of instructions that make a computer do something useful. It's "soft" in contrast with "hard" hardware because it's intangible, not easy to put your hands on. Hardware is tangible: if you drop a laptop on your foot, you'll notice. Not true for software.

In the next few chapters we're going to talk about software: how to tell a computer what to do. Chapter 4 is a discussion of software in the abstract, focusing on algorithms, which are in effect idealized programs for focused tasks. Chapter 5 discusses programming and programming languages, which we use to express sequences of computational steps. Chapter 6 describes the major kinds of software systems that we all use, whether we know it or not. The final chapter of this section, Chapter 7, is a brief introduction to programming in two of today's most popular languages, JavaScript and Python.

Something to keep in mind as we go: modern technological systems increasingly use general-purpose hardware—processors, memory, and connections to the environment—and create specific behaviors by software. The conventional wisdom is that software is cheaper, more flexible, and easier to change than hardware is, especially once some device has left the factory. For example, if a computer controls how power and brakes are applied in a car, apparently different functions like anti-lock braking and electronic stability control are software features.

Trains, boats, and airplanes are also increasingly reliant on software. Unfortunately, it's not always simple to use software to change physical behavior. Airplane software was much in the news in the aftermath of two fatal crashes of Boeing's 737 MAX in October 2018 and March 2019, in which 346 people died.

Boeing started making 737s in 1967, and the plane evolved steadily over the years. The 737 MAX, which went into service in 2017, was a major modification with bigger and more efficient engines.

The new engines caused the plane to have significantly different flying characteristics. Rather than making aerodynamic modifications to keep its behavior close to earlier models, Boeing developed an automated flight control software system called the Maneuvering Characteristics Augmentation System, or MCAS. The intent of MCAS was that the MAX would fly the same as any other 737, and thus would not need recertification, nor would pilots need to be retrained, both expensive processes: software would make the new plane just like the older ones.

To over-simplify a complicated situation, the heavier and repositioned engines changed the MAX's flying characteristics. In some circumstances, when MCAS thought that the plane's nose was too high, it interpreted this as a potential stall and pushed the nose down. It based its decision on a single potentially faulty input sensor, even though the plane had two sensors. When pilots tried to pull the nose up, MCAS overrode them. The result was a series of up and down oscillations that eventually caused the fatal crashes. To make the story even worse, Boeing had not revealed the existence of MCAS, so pilots were not aware of the potential problem and thus were not properly trained to deal with it.

Soon after the second fatal crash, aviation authorities around the world grounded the MAX. Boeing's reputation was badly tarnished, and estimates of its losses run to more than $20 billion. In late November 2020, the US Federal Aviation Authority cleared the MAX to fly again after changes are made in pilot training and in the plane itself, but it's not clear when it will return to regular service.

Computers are central to critical systems, and software controls them. Self-driving cars, or merely the assists provided by modern cars, are controlled by software. As a simple example, my Subaru Forester has two cameras that look out the front windshield; it uses computer vision to warn me if I change lanes without signaling or when a car or a person appears to be too close. It's often wrong and the frequent false positives are more distracting than helpful, but it has saved me a few times.

Medical imaging systems use computers to control signals and form images for doctors to interpret, and film has been replaced by digital images. It's also true of infrastructure like the air traffic control system, navigational aids, the power grid and the telephone network. Computer-based voting machines have had serious flaws. In early 2020, vote tabulation for the Iowa Democratic primaries was a computer-system fiasco that took days to fix. Internet voting, a popular idea during the Covid-19 pandemic, is much riskier than election officials acknowledge; it's very difficult to create a system that lets people vote securely while preserving the secrecy of how they voted.

Military systems for both weapons and logistics are entirely dependent on computers, as are the world's financial systems. Cyber warfare and espionage are real threats. The Stuxnet worm of 2010 destroyed uranium-enrichment centrifuges in Iran. A large power outage in Ukraine in December 2015 was caused by malware of Russian origin, though the Russian government denies involvement. Two years later, a second set of attacks using ransomware called Petya interfered with a variety of Ukrainian services. The 2017 ransomware attack known as WannaCry caused billions of dollars of damages all over the world; the US government formally accused North Korea of being responsible. In July 2020, a Russian cyber espionage group was accused by several countries of trying to steal information about potential Covid-19 vaccines.

State-sponsored and criminal attacks are quite possible on a wide variety of targets. If our software is not reliable and robust, we're in trouble and it's only going to get worse as our dependency grows. As we'll see, it's hard to write software that is completely reliable. Any error or oversight in logic or implementation can cause a program to behave incorrectly, and even if that doesn't happen in normal use, it might leave an opening for an attacker.

4

Algorithms

The Feynman Algorithm:
1. Write down the problem.
2. Think real hard.
3. Write down the solution.

Attributed to physicist Murray Gell-Mann, 1992.

A popular metaphor for explaining software is to compare it to recipes for cooking. A recipe for some dish lists the ingredients needed, the sequence of operations that the cook has to perform, and the expected outcome. By analogy, a program for some task needs data to operate on and spells out what to do to the data. Real recipes are much more vague and ambiguous than programs could ever be, however, so the analogy isn't terribly good. For instance, a chocolate cake recipe says "Bake in the oven for 30 minutes or until set. Test by placing the flat of your hand gently on the surface." What should the tester be looking for—wiggling, resistance, or something else? How gentle is "gently"? Should the baking time be at least 30 minutes or no longer than 30 minutes?

Tax forms are a better metaphor: they spell out in painful detail what to do ("Subtract line 30 from line 29. If zero or less, enter 0. Multiply line 31 by 25%, ..."). The analogy is still imperfect but tax forms capture the computational aspects much better than recipes do: arithmetic is required, data values are copied from one place to another, conditions are tested, and subsequent computations depend on the results of earlier ones.

For taxes especially, the process should be complete—it should always produce a result, the amount of tax due, no matter what the situation. It should be unambiguous—anyone starting from the same initial data should arrive at the same final answer. And it should stop after a finite amount of time. Speaking from personal experience, these are all idealizations, because terminology is not always clear, instructions are more ambiguous than the tax authorities care to admit, and frequently it's unclear what data values to use.

An *algorithm* is the computer science version of a careful, precise, unambiguous recipe or tax form, a sequence of steps that is guaranteed to compute a result correctly. Each step is expressed in terms of basic operations whose meaning is completely specified, for example "add two integers." There's no ambiguity about what anything means. The nature of the input data is given. All possible situations are covered; the algorithm never encounters a situation where it doesn't know what to do next. When in a pedantic mood, computer scientists usually add one more condition: the algorithm has to stop eventually. By that standard, the classic shampoo instruction to "Lather, Rinse, Repeat" is not an algorithm.

The design, analysis and implementation of efficient algorithms is a core part of academic computer science, and there are real-world algorithms of great importance. I'm not going to try to explain or express algorithms precisely, but I do want to get across the idea of specifying a sequence of operations with sufficient detail and precision that there is no doubt about what the operations mean and how to perform them, even if they are being performed by an entity with no intelligence or imagination. We will also discuss algorithmic efficiency, that is, how computation time depends on the amount of data to be processed. We'll do this for a handful of basic algorithms that are familiar and readily understood.

You don't have to follow all the details or the occasional formulas in this chapter, but the ideas are important.

4.1 Linear Algorithms

Suppose we want to find out who is the tallest person in the room. We could just look around and make a guess, but an algorithm must spell out the steps so precisely that even a dumb computer can follow them. The basic approach is to ask each person in turn about their height, and keep track of the tallest person seen so far. Thus we might ask each person in turn: "John, how tall are you? Mary, how tall are you?" and so on. If John is the first person we ask, then so far he's the tallest. If Mary is taller, she's now the tallest; otherwise, John retains his position. Either way, we ask the third person. At the end of the process, after each person has been asked, we know who the tallest person is and his or her height. Obvious variations would find the richest person, or the person whose name comes first in the alphabet, or the person whose birthday is closest to the end of the year.

There are complications. How do we handle duplicates, for instance two or more people who are exactly the same height? We have to choose whether to report the first or the last or a random one, or perhaps all of them. Notice that finding the largest group of people who are equally tall is a significantly harder problem, since it means we have to remember the names of all people who are the same height: we won't know until the end of the input who is on the final list. This example involves *data structures*—how to represent the information needed during a computation— which are an important consideration for many algorithms, though we won't talk much about them here.

What if we want to compute the average height? We could ask each person for his or her height, add up the values as they are obtained (perhaps using the Toy program for adding up a sequence of numbers), and at the end divide the sum by the number of people. If there's a list of N heights written on a piece of paper, we might express this example more "algorithmically" like this:

```
set sum to 0
for each height on the list     average hight
    add the height to sum
set average to sum / N
```

We have to be more careful if we're asking a computer to do the job, however. What happens, for example, if there are no numbers on the piece of paper? This isn't a problem if a person is doing the work, since we realize that there's nothing to do. A computer, by contrast, has to be told to test for this possibility and how to act if it occurs. If the test isn't made, the result will be an attempt to divide the sum by zero, which is an undefined operation. Algorithms and computers have to handle all possible situations. If you've ever received a check made out for "0 dollars and 00 cents" or a bill telling you to pay a balance due of zero, you've seen an example of failing to test all cases properly.

What if we don't know ahead of time how many data items there will be, as is usually the case? Then we can count the items as we compute the sum:

```
set sum to 0                            better average hight
set N to 0
repeat these two steps for each height:
    add the next height to sum
    add 1 to N
if N is greater than 0
    set average to sum divided by N
otherwise
    report that no heights were given
```

This shows one way to handle the problem of potential division by zero, by testing for the awkward case explicitly.

One crucial property of algorithms is how efficiently they operate—are they fast or slow, and how long are they likely to take to handle a given amount of data? For the examples given above, the number of steps to perform, or how long it might take a computer to do the job, is directly proportional to the amount of data that must be processed. If there are twice as many people in the room, it will take twice as long to find the tallest or compute the average height, and if there are ten times as many, it will take ten times as long. When the computation time is directly or linearly proportional to the amount of data, the algorithm is called *linear-time* or just *linear*. If we were to plot running time against the number of data items, the graph would be a straight line pointing upward and to the right. Many of the algorithms that we encounter in day-to-day life are linear, since they involve doing the same basic operation or operations on some data, and more data means more work in direct proportion.

Many linear algorithms take the same basic form. There may be some initialization, like setting the running sum to zero or the largest height to a small value. Then each item is examined in turn and a simple computation is done on it—counting it, comparing it to a previous value, transforming it in a simple way, perhaps printing it. At the end, some step may be needed to finish the job, like computing the average or printing the sum or the largest height. If the operation on each item takes about the same amount of time, then the total time is proportional to the number of items.

4.2 Binary Search

Can we ever do any better than linear time? Suppose we have a bunch of names and telephone numbers in a printed list or a stack of business cards. If the names are in no particular order and we want to find Mike Smith's number, then we have to look at all the cards until we find his name, or fail to find it because it's not there at all. If the names are in alphabetical order, however, we can do better.

Think about how we look up a name in an old-fashioned paper phone book. We start approximately in the middle. If the name we're looking for is earlier in the alphabet than names on the middle page, we can completely ignore the last half of the book and look next at the middle of the first half (one quarter of the way into the original); otherwise, we ignore the first half and check the middle of the last half (three quarters of the way in). Because the names are in alphabetical order, at each step we know which half to look in next. Eventually we get down to the point where either the name we're looking for has been found, or we know for sure that it's not there at all.

This search algorithm is called *binary search*, because each check or comparison divides the items into two groups, one of which can be eliminated from further consideration. It's an example of a general strategy called *divide and conquer*. How fast is it? At each step, half the remaining items are eliminated, so the number of steps is the number of times we can divide the original size by 2 before we get down to a single item.

Suppose we start with 1,024 names, a number chosen to make the arithmetic easy. With one comparison, we can eliminate 512 names. Another comparison gets us down to 256, then 128, then 64, then 32, then 16, 8, 4, 2, and finally 1. That's 10 comparisons. It's clearly not a coincidence that 2^{10} is 1,024. The number of comparisons is the power of 2 that brings us to the original number; running the sequence up from 1 to 2 to 4 ... to 1,024, we're multiplying by 2 each time.

If you remember logarithms from school (not many people do—who would have thought?), you may recall that the logarithm of a number is the power to which you raise the base (2 in this case) to get the number. So the log (base 2) of 1,024 is 10, since 2^{10} is 1,024. For our purposes, the log is the number of times you have to divide a number by 2 to get down to 1 or, equivalently, the number of times you have to multiply 2 by itself to get up to the number. In this book, *log* will always mean base 2. We don't need precision or fractions; ballpark figures and integer values are good enough, which is a real simplification.

The important thing about binary search is that the amount of work to be done grows slowly with the amount of data. If there are 1,000 names in alphabetical order, we have to look at 10 names to find a specific one. If there are 2,000 names, we only have to look at 11 names because the first name we look at immediately eliminates 1,000 of the 2,000, bringing us back to where we have to look at 1,000 (10 tests). If there are a million names, that's 1,000 times 1,000. The first 10 tests get us back down to 1,000, and another 10 get us down to 1, which is 20 tests in all. A million is

10^6, which is close to 2^{20}, so the log (base 2) of a million is about 20.

From this, you should be able to see that looking up a name in a directory of a billion names (the whole earth phone book, nearly) would only take 30 name comparisons, since a billion is about 2^{30}. That's why we say the amount of work grows slowly with the amount of data—a thousand times more data requires only 10 more steps.

As a quick validation, I decided to search for my friend Harry Lewis in an old Harvard paper telephone directory, which at the time had about 20,000 names in 224 pages. (Of course, paper phone books have long since disappeared, so I can't repeat the experiment today.) I began at page 112, finding the name Lawrence. "Lewis" comes later than that; it's in the second half, so I next tried page 168, halfway between 112 and 224, to find Rivera. "Lewis" is before that, so I tried 140 (halfway between 112 and 168), leading to Morita. Next to 126 (halfway between 112 and 140) to find Mark. The next try was 119 (Little), then 115 (Leitner), then 117 (Li), to 116. There were about 90 names on this page, so another 7 comparisons, all on the same page, found Harry among a dozen other Lewises. This experiment took 14 tests in all, which is about what would be expected, since 20,000 lies between 2^{14} (16,384) and 2^{15} (32,768).

This kind of binary division shows up in real-world settings like the knockout tournaments used in many sports. The tournament starts with a large number of competitors, for example 128 for men's singles tennis at Wimbledon. Each round eliminates half the contenders, leaving one pair in the final round, from which a single winner emerges. Not by coincidence, 128 is a power of two (2^7), so Wimbledon has seven rounds. One could imagine a worldwide knockout tournament; even with seven billion participants, it would only take 33 rounds to determine the winner. If you recall the discussion of powers of two and ten in Chapter 2, you can verify this with easy mental arithmetic.

4.3 Sorting

But how do we get those names into alphabetical order in the first place? Without that preliminary step, we can't use binary search. This brings us to another fundamental algorithmic problem, *sorting*, which puts things into order so that subsequent searches can run quickly.

Suppose we want to sort some names into alphabetical order so we can later search them efficiently with binary search. One algorithm is called *selection sort*, since it keeps selecting the next name from among the ones not yet in order. It's based on the technique for finding the tallest person in the room that we saw earlier.

Let's illustrate by sorting these 16 familiar names alphabetically:

  Intel Facebook Zillow Yahoo Pinterest Twitter Verizon Bing
  Apple Google Microsoft Sony PayPal Skype IBM Ebay

Start at the beginning. Intel is first, so it's alphabetically first so far. Compare that to the next name, Facebook. Facebook comes earlier in the alphabet, so it temporarily becomes the new first name. Zillow is not before Facebook, nor are any other names until Bing, which replaces Facebook; Bing is in turn replaced by Apple.

We examine the rest of the names but none of them precedes Apple, so Apple is truly first in the list. We move Apple to the front and leave the rest of the names as they were. The list now looks like this:

Apple

Intel Facebook Zillow Yahoo Pinterest Twitter Verizon Bing
Google Microsoft Sony PayPal Skype IBM Ebay

Now we repeat the process to find the second name, starting with Intel, which is the first name in the unsorted group. Again Facebook replaces it, then Bing becomes the first element. After the second pass is complete, the result is:

Apple Bing

Intel Facebook Zillow Yahoo Pinterest Twitter Verizon
Google Microsoft Sony PayPal Skype IBM Ebay

After 14 more steps this algorithm produces a completely sorted list.

How much work does it do? It makes repeated passes through the remaining items, each time finding the next name in alphabetical order. With 16 names, identifying the first name requires looking at 16 names. Finding the second name takes 15 steps, finding the third takes 14 steps, and so on. In the end, we have to look at 16+15+14+...+3+2+1 names, 136 in all. Of course a clever sorting algorithm could discover that the names are already in order, but computer scientists who study algorithms are pessimists and assume the worst case, where there are no shortcuts and all the work has to be done.

The number of passes through the names is directly proportional to the original number of items (16 for our example, or N in general). Each pass has to deal with one less item, however, so the amount of work in the general case is

$$N + (N - 1) + (N - 2) + (N - 3) + \cdots + 2 + 1$$

This series adds up to $N \times (N + 1)/2$ (most easily seen by adding the numbers in pairs from each end), which is $N^2/2 + N/2$. Ignoring the division by 2, the work is proportional to $N^2 + N$. As N gets bigger, N^2 quickly becomes much bigger than N (for instance, if N is a thousand, N^2 is a million), so the effect is that the amount of work is approximately proportional to N^2 or the square of N, a growth rate called *quadratic*. Quadratic is worse than linear; in fact, it's much worse. If there are twice as many items to sort, it will take four times as long; if there are ten times as many items, it will take one hundred times as long; and if there are a thousand times as many items, it will take a million times as long! This is not good.

Fortunately, it is possible to sort much more quickly. Let's take a look at one clever way, an algorithm called *Quicksort*, which was invented by the English computer scientist Tony Hoare around 1959. (Hoare won the Turing Award in 1980 for multiple contributions, including Quicksort.) It's an elegant algorithm, and a great example of divide and conquer.

Here are the unsorted names again:

Intel Facebook Zillow Yahoo Pinterest Twitter Verizon Bing
Apple Google Microsoft Sony PayPal Skype IBM Ebay

To sort the names with a simplified version of Quicksort, first go through the names once, putting the ones that begin with A through M in one pile and N through Z in another. That produces two piles, each having about half the names. This assumes that the distribution of names isn't badly skewed, so that approximately half the names at each stage fall into each pile. In our case, the two resulting piles each have eight names:

Intel Facebook Bing Apple Google Microsoft IBM Ebay
Zillow Yahoo Pinterest Twitter Verizon Sony PayPal Skype

Now go through the A–M pile, putting A through F in one pile and G through M in another; go through the N–Z pile, putting N–S in one pile and T–Z in another. At this point, we have made two passes through the names, and have four piles, each with about one quarter of the names:

Facebook Bing Apple Ebay
Intel Google Microsoft IBM
Pinterest Sony PayPal Skype
Zillow Yahoo Twitter Verizon

The next pass goes through each of those piles, splitting the A–F pile into ABC and DEF and G–M into GHIJ and KLM, and the same for N–S and T–Z; at this point we have eight piles with about two names in each:

Bing Apple
Facebook Ebay
Intel Google IBM
Microsoft
Pinterest PayPal
Sony Skype
Twitter Verizon
Zillow Yahoo

Of course, eventually we will have to look at more than just the first letter of the names, for example to put IBM ahead of Intel and Skype ahead of Sony. But after one or two more passes, we'll have 16 piles of one name each, and the names will be in alphabetical order.

How much work was that? We looked at each of the 16 names in each pass. If the split were perfect each time, the piles would have 8 names, then 4, then 2, then 1. The number of passes is the number of times we divide 16 by 2 to get down to 1. That's the log base 2 of 16, which is 4. Thus the amount of work is $16 \log_2 16$ for 16 names. If we make four passes through the data, that's 64 operations, compared to 136 for selection sort. That's for 16 names; when there are more, Quicksort's advantage will be much greater, as can be seen in Figure 4.1.

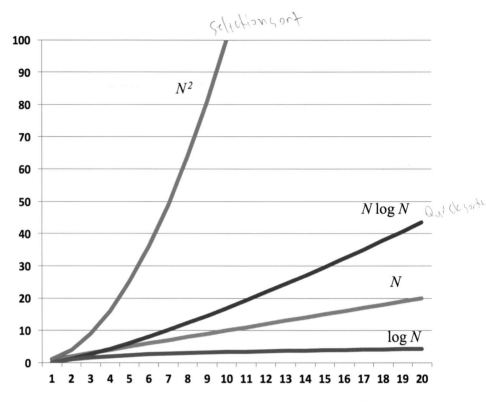

Selection sort

Quick sort

Figure 4.1: Growth of $\log N$, N, $N \log N$ and N^2.

Catch

This algorithm will always sort the data, but it is only efficient if each division divides the group into piles of about the same size. On real data, Quicksort has to guess the middle data value so it can split into two approximately equal-size groups each time; in practice this can be estimated well enough by sampling a few items. In general, Quicksort takes about $N \log N$ operations to sort N items; the amount of work is proportional to $N \times \log N$. That's worse than linear, but not terribly so, and it's enormously better than quadratic or N^2 when N is at all large.

The graph in Figure 4.1 shows how $\log N$, N, $N \log N$ and N^2 grow as the amount of data grows; it plots them for 20 values, though only 10 for quadratic, which would otherwise be through the roof.

As an experiment, I generated 10 million random 9-digit numbers, analogous to US Social Security numbers, and timed how long it took to sort groups of various sizes with selection sort (N^2 or quadratic) and with Quicksort ($N \log N$); the results are in Figure 4.2. The dashes in the table are cases I did not run.

It's hard to accurately measure programs that run for short times, so take these numbers with a large grain of salt. Nevertheless, you can see roughly the expected $N \log N$ growth of run time for Quicksort, and you can also see that selection sort might be feasible for up to say 10,000 items, though far from competitive; at every stage, it's hopelessly outclassed by Quicksort.

You might also have noticed that the selection sort time for 100,000 items is nearly 200 times larger than for 10,000, rather than the expected 100 times. This is

Number of numbers (N)	Selection sort time (sec)	Quicksort time (sec)
1,000	0.047	-
10,000	4.15	0.025
100,000	771	0.23
1,000,000	-	3.07
10,000,000	-	39.9

Figure 4.2: Comparison of sorting times.

likely a caching effect—the numbers don't all fit in the cache and thus sorting is slower. It's a good illustration of the difference between an abstraction of computational effort and the reality of a concrete computation by an actual program.

4.4 Hard Problems and Complexity

We've now studied several points along a spectrum of algorithmic "complexity" or running time. At one end is $\log N$, as seen in binary search, where the amount of work grows slowly as the amount of data increases. The most common case is linear, or plain N, where the work is directly proportional to the amount of data. There's $N \log N$, as in Quicksort; that's worse (grows faster) than N, but still eminently practical for large values of N because the log factor grows so slowly. And there's N^2, or quadratic, which grows quickly enough that it's somewhere between painful and impractical.

There are plenty of other complexity possibilities, some easy to understand, like cubic, or N^3, which is worse than quadratic but the same idea, and others so esoteric that only specialists could care about them. One more is worth knowing, since it occurs in practice, it's especially bad, and it's important. *Exponential* complexity grows like 2^N (which is not the same as N^2). In an exponential algorithm, the amount of work grows exceptionally rapidly: adding one more item *doubles* the amount of work to be done. In a sense, exponential is at the opposite end from a $\log N$ algorithm, where doubling the number of items adds only one more step.

Exponential algorithms arise in situations where in effect we must try all the possibilities one by one. Fortunately, there is a silver lining to the existence of problems that require exponential algorithms. Some algorithms, notably in cryptography, are based on the exponential difficulty of performing a specific computational task. For such algorithms, one chooses N large enough that it's not computationally feasible to solve the problem directly without knowing a secret shortcut—it would take far too long—and that provides the protection from adversaries. We'll take a look at cryptography in Chapter 13.

By now you should have an intuitive understanding that some problems are easy to deal with, while others seem harder. It's possible to make this distinction more precise. "Easy" problems are "polynomial" in their complexity; that is, their running time is expressed as some polynomial like N^2, though if the exponent is more than 2, they are likely to be challenging. (Don't worry if you've forgotten what a polynomial is—here, think of it as an expression with only integer powers of a variable, like N^2

or N^3.) Computer scientists call this class of problems "P", for "polynomial," because they can be solved in polynomial time.

A large number of problems that occur in practice or as the essence of practical problems seem to require exponential algorithms to solve; that is, we know of no polynomial algorithm. These problems are called "NP" problems. NP problems have the property that we can't find a solution quickly but we can verify quickly that a proposed solution is correct. NP stands for "nondeterministic polynomial," which informally means that they can be solved in polynomial time by an algorithm that always guesses right when it has to make a choice. In real life, nothing is lucky enough to always choose correctly, so this is a theoretical idea only.

Many NP problems are quite technical, but one is easy to explain and its practical applications can be imagined. In the *Traveling Salesman Problem (TSP)*, a salesman has to start from his or her home city, visit a number of other specific cities in any order, and then return home. The goal is to visit each city exactly once (no repeats) and to travel the minimum total distance. This captures the idea of efficiently routing school buses or garbage trucks; when I worked on it long ago, it was used for tasks as diverse as planning how to drill holes in circuit boards and sending boats to pick up water samples at specific places in the Gulf of Mexico.

Figure 4.3 shows a randomly generated 10-city problem, with a solution found by the intuitively appealing "nearest neighbor" heuristic: start at some city and at each city, go next to the nearest unvisited city. The length of this tour is 12.92. Note that different starting cities can lead to different tours; the tour in Figure 4.3 is the shortest of those.

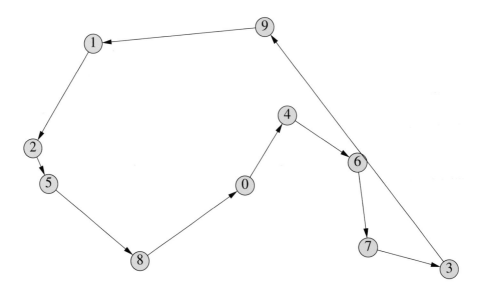

Figure 4.3: Nearest-neighbor solution to a 10-city TSP (length 12.92).

The Traveling Salesman Problem was first described in the 1800s and has been the subject of intense study for many years. Although we're better now at solving larger instances, techniques for finding the best solution still amount to clever variations of trying all possible routes. For comparison, Figure 4.4 is the shortest tour, found by exhaustive search of all 180,000 tours; its length is 11.86, about 8 percent shorter than the best nearest-neighbor tour.

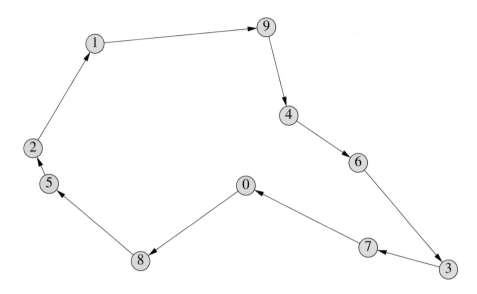

Figure 4.4:   Best solution to the 10-city TSP (length 11.86).

The same is true for a wide variety of problems: we don't have good ways to solve them efficiently other than by exhaustive search of all possible solutions. For people who study algorithms, this is frustrating. We don't know whether these problems are intrinsically hard or we're just dumb and haven't figured out how to deal with them yet, though the betting today heavily favors the "intrinsically hard" side.

A remarkable mathematical result by Stephen Cook in 1971 showed that many of these problems are equivalent, in the sense that if we could find a polynomial-time algorithm (that is, something like N^2) for any one of them, that would enable us to find polynomial-time algorithms for all of them. Cook won the 1982 Turing Award for this work.

In 2000, the Clay Mathematics Institute offered a prize of one million dollars each for the solution of seven unsolved problems. One of these questions is to determine whether P equals NP, that is, whether the hard problems are really the same class as the easy problems. The Poincaré Conjecture, another problem on the list, dates from the early 1900s. It was resolved by the Russian mathematician Grigori Perelman and the prize was awarded in 2010, though Perelman declined to accept it. Only six problems are left—better hurry before someone beats you to it!

There are a couple of things to keep in mind about this kind of complexity. Although the P=NP question is important, it is more a theoretical issue than a practical one. Most complexity results stated by computer scientists are for the *worst case*. That is, some problem instances will require the maximum time to compute the answer, but not all instances need be that hard. They are also *asymptotic* measures that only apply for large values of N. In real life, N might be small enough that the asymptotic behavior doesn't matter. For instance, if you're only sorting a few dozen or even a few hundred items, selection sort might be fast enough even though its complexity is quadratic and thus asymptotically much worse than Quicksort's $N \log N$. If you're only visiting 10 cities, it's feasible to try all possible routes, but that would likely be infeasible for 100 cities and definitely impossible for 1,000. Finally, in most real situations an approximate solution is probably good enough; there's no need for an absolutely optimal solution.

On the flip side, some important applications, like cryptographic systems, are based on the belief that a particular problem truly is difficult, so the discovery of an attack, however impractical it might be in the short run, could be significant.

4.5 Summary

Computer science as a field has spent years refining the notion of "how fast can we compute"; the idea of expressing running time in terms of the amount of data like N, $\log N$, N^2, or $N \log N$ is the distillation of that thinking. It ignores concerns like whether one computer is faster than another, or whether you are a better programmer than I am. It does capture the complexity of the underlying problem or algorithm, however, and for that reason, it's a good way to make comparisons and to reason about whether some computation is likely to be feasible or not. (The intrinsic complexity of a problem and the complexity of an algorithm to solve it need not be the same. For example, sorting is an $N \log N$ problem, for which Quicksort is an $N \log N$ algorithm, but selection sort is an N^2 algorithm.)

The study of algorithms and complexity is a major part of computer science, as both theory and practice. We're interested in what can be computed and what can't, and how to compute fast and without using more memory than necessary or perhaps by trading off speed against memory. We look for fundamentally new and better ways to compute; Quicksort is a nice example of that, though from long ago.

Many algorithms are more specialized and complicated than the basic searching and sorting that we've talked about. For example, compression algorithms attempt to reduce the amount of memory occupied by text, music (MP3, AAC), images and pictures (PNG, JPEG) and movies (MPEG). Error detection and correction algorithms are also important. Data is subject to potential damage as it is stored and transmitted, for example over noisy wireless channels or CDs that have been scratched; algorithms that add controlled redundancy to the data make it possible to detect and even correct some kinds of errors. We'll come back to these algorithms in Chapter 8 since they have implications when we talk about communications networks.

Cryptography, the art of sending secret messages so they can be read only by the intended recipients, depends strongly on algorithms. We'll discuss cryptography in Chapter 13, since it's highly relevant when computers exchange private information in a secure fashion.

Search engines like Bing and Google are yet another place where algorithms are crucial. In principle, much of what a search engine does is simple: collect some web pages, organize the information to make it easy to search, and then search it efficiently. The problem is scale. When there are billions of web pages and billions of queries every day, even $N \log N$ isn't good enough, and much algorithmic and programming cleverness goes into making search engines run fast enough to keep up with the growth of the web and our interest in finding things on it. We'll talk more about search engines in Chapter 11.

Algorithms are also at the heart of services like speech understanding, face and image recognition, machine translation of languages, and so on. These all depend on having lots of data that can be mined for relevant features, so the algorithms have to be linear or better and generally have to be parallelizable so separate pieces can run on multiple processors simultaneously. More on this in Chapter 12.

5

Programming and
Programming Languages

"The realization came over me with full force that a good part of the remainder of my life was going to be spent in finding errors in my own programs."

Maurice Wilkes, *Memoirs of a Computer Pioneer*, 1985.

So far we have talked about algorithms, which are abstract or idealized process descriptions that ignore details and practicalities. An algorithm is a precise and unambiguous recipe. It's expressed in terms of a fixed set of basic operations whose meanings are completely known and specified. It spells out a sequence of steps using those operations, with all possible situations covered, and it's guaranteed to stop eventually.

By contrast, a *program* is anything but abstract—it's a concrete statement of every step that a real computer must perform to accomplish a task. The distinction between an algorithm and a program is like the difference between a blueprint and a building; one is an idealization and the other is the real thing.

One way to view a program is as one or more algorithms expressed in a form that a computer can process directly. A program has to worry about practical problems like insufficient memory, limited processor speed, invalid or even malicious input data, faulty hardware, broken network connections, and (in the background and often exacerbating the other problems) human frailty. So if an algorithm is an idealized recipe, a program is the detailed set of instructions for a cooking robot preparing a month of meals for an army while under enemy attack.

We can only get so far with metaphors, of course, so we're going to talk about real programming enough that you can understand what's going on, though not enough to make you into a professional programmer. Programming can be hard—there are many details to get right and tiny slip-ups can lead to large errors—but it's not impossible and it can be a great deal of fun, as well as a marketable skill.

There aren't enough programmers in the world to do the amount of programming involved in making computers do everything we want or need. So one continuing theme in computing has been to enlist computers to handle more and more of the details of programming. This leads to a discussion of programming languages: languages that let us express the computational steps needed to perform some task in a form that's more or less natural for humans.

It's also hard to manage the resources of a computer, especially given the complexity of modern hardware. So we also use the computer to control its own operations, which leads us to operating systems. Programming and programming languages are the topic of this chapter, and software systems, especially operating systems, will come in the next. Chapter 7 takes a more detailed look at two important languages, JavaScript and Python.

You can certainly skip the syntactic details in the programming examples in this chapter, but it's worth looking at similarities and differences in how computations are expressed.

5.1 Assembly Language

For the first truly programmable electronic computers, programming was a laborious process. Programmers had to convert instructions and data into binary numbers, make these numbers machine-readable by punching holes into cards or paper tape, then load them into the computer memory. Programming at this level was incredibly difficult even for tiny programs—hard to get right in the first place, and hard to change if a mistake was found or instructions and data had to be changed or added.

Maurice Wilkes's comment in the epigraph above hints at the challenge. Wilkes was the designer and implementer of EDSAC, one of the first stored-program computers; it became operational in 1949. He won the Turing Award in 1967 for his contributions, and was knighted in 2000.

Early in the 1950s, programs were created to handle some of the straightforward clerical chores, so that programmers could use meaningful words for instructions (ADD instead of 5, for example) and names for specific memory locations (Sum instead of 14). This powerful idea—a program to manipulate another program—has been at the heart of most significant advances in software.

The program that does this specific manipulation is called an *assembler* because originally it also assembled any necessary parts of a program that had been written earlier by other programmers. The language is called *assembly language* and programming at this level is called *assembly language programming*. The language that we used to describe and program the Toy computer in Chapter 3 is an assembly language. Assemblers make it much easier to modify a program, because when the programmer adds or removes instructions, the assembler keeps track of where each instruction and data value will be located in memory, rather than requiring the programmer to do the bookkeeping by hand.

An assembly language for a particular processor architecture is specific to that architecture; it usually matches the instructions of the processor one for one and it knows the specific way that instructions are encoded in binary, how information is placed in memory, and so on. This means that a program written in the assembly language of one particular kind of processor, say an Intel processor in a Mac or PC, will be different from an assembly language program for the same task for a different

CPU, like the ARM processor in a cell phone. If one wants to convert an assembly language program from one of those processors to the other, the program must be totally rewritten.

To make this concrete, in the Toy computer it takes three instructions to add two numbers and store the result in a memory location:

```
LOAD   X
ADD    Y
STORE  Z
```

This would be similar in a variety of current processors. In a CPU with a different instruction repertoire, however, this computation might be accomplished with a sequence of two instructions that access memory locations without using an accumulator:

```
COPY X, Z
ADD   Y, Z
```

To convert a Toy program to run on the second computer, a programmer would have to be intimately familiar with both processors and meticulous in converting from one instruction set to the other. It's hard work.

5.2 High-Level Languages

During the late 1950s and early 1960s, another step was taken towards getting the computer to do more for programmers, arguably the most important step in the history of programming. This was the development of *high-level programming languages* that were independent of any particular processor architecture. High-level languages make it possible to express computations in terms that are closer to the way a person might express them.

Code written in the high-level language is converted by a translator program into instructions in the assembly language of a specific target processor, which in turn are converted by the assembler into bits to be loaded into the memory and executed. The translator is usually called a *compiler*, another historical term that doesn't convey much insight or intuition.

In a typical high-level language, the computation above that adds two numbers X and Y and stores the result as a third, Z, would be expressed as

```
Z = X + Y
```

This means "get the values from the memory locations called X and Y, add them together, and store the result in the memory location called Z." The operator "=" means "replace" or "store," not "equal to."

A compiler for the Toy would convert this into the sequence of three instructions, while a compiler for the other computer would convert it into two instructions. The respective assemblers would then be responsible for converting their assembly language instructions into the actual bit patterns of real instructions, as well as setting aside memory locations for the quantities X, Y and Z. The resulting bit patterns would almost certainly be different for the two computers.

This process is illustrated in Figure 5.1, which shows the same input expression going through two different compilers and their corresponding assemblers, to produce two different sequences of instructions.

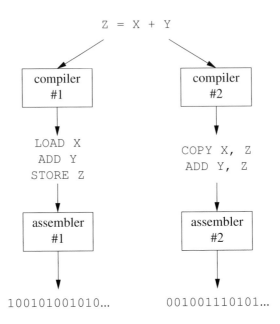

Figure 5.1: The compilation process with two compilers.

As a practical matter, a compiler will likely be divided internally into a "front end" that processes programs in the high-level programming language into some intermediate form, and several "back ends," each one of which converts the common intermediate form into the assembly language for its specific architecture. This organization is simpler than having multiple completely independent compilers.

High-level languages have great advantages over assembly language. Because high-level languages are closer to the way that people think, they are easier to learn and use; one doesn't need to know anything about the instruction repertoire of any particular processor to program effectively in a high-level language. Thus they make it possible for more people to program computers and to program them more quickly.

Second, a program in a high-level language is independent of any particular architecture, so the same program can be run on different architectures, usually without any changes at all, merely by compiling it with a different compiler, as in Figure 5.1. The program is written only once but can be run on different computers. This makes it possible to amortize development costs over multiple kinds of computers, even ones that don't yet exist.

The compilation step also provides a preliminary check for some kinds of gross errors—misspellings, syntax errors like unbalanced parentheses, operations on undefined quantities, and so on—that the programmer must correct before an executable program can be produced. Some such errors are hard to detect in assembly language programs, where any sequence of instructions must be presumed legal. (Of course a syntactically correct program can still be full of errors undetectable by a compiler.) It's hard to overstate the importance of high-level languages.

I'm going to show the same program in six of the most important high-level programming languages—Fortran, C, C++, Java, JavaScript, and Python—so you can get a sense of their similarities and differences. Each program does the same thing as the program that we wrote for the Toy in Chapter 3. It adds up a sequence of integers; when a zero value is read, it prints the sum and stops. The programs all have the same structure: name the quantities that the program uses, initialize the running sum to zero, read numbers and add them to the running sum until zero is encountered, then print the sum. Don't worry about syntactic details; this is mostly to give you an impression of what the languages look like. I've tried to keep the examples as similar as possible, even though that might not be the best way to write them individually.

The first high-level languages concentrated on specific application domains. One of the earliest languages was called FORTRAN, a name derived from "Formula Translation" and written today as "Fortran." Fortran was developed by an IBM team led by John Backus, and was very successful for expressing computations in science and engineering. Many scientists and engineers (including me) learned Fortran as their first programming language. Fortran is alive and well today; it has gone through several evolutionary stages since 1958, but is recognizably the same language. Backus received the Turing Award in 1977 in part for his work on Fortran.

Figure 5.2 shows a Fortran program to add up a series of numbers.

```
      integer num, sum
      sum = 0
   10 read(5,*) num
      if (num .eq. 0) goto 20
      sum = sum + num
      goto 10
   20 write(6,*) sum
      stop
      end
```

Figure 5.2: Fortran program to add up numbers.

This is written in Fortran 77; it would look a bit different in an earlier version or a later one like Fortran 2018, the most recent. You can imagine how to translate both the arithmetic expressions and the sequencing of operations into Toy assembly language. The `read` and `write` operations obviously correspond to `GET` and `PRINT`, and the fourth line is clearly an `IFZERO` test.

A second major high-level language of the late 1950s was COBOL (Common Business Oriented Language), which was strongly influenced by the work of Grace Hopper on high-level alternatives to assembly language. Hopper worked with Howard Aiken on the Harvard Mark I and II, early mechanical computers, and then on the UNIVAC I. She was one of the first people to see the potential of high-level languages and compilers. COBOL was specifically aimed at business data processing, with language features to make it easy to express the kinds of data structures and computations that go into managing inventories, preparing invoices, computing payrolls, and the like. COBOL too lives on, much changed but still recognizable. There are a lot of legacy COBOL programs, but not many COBOL programmers. In 2020, the state of New Jersey discovered that their ancient programs for processing unemployment claims couldn't cope with the increased volume caused by Covid-19, but the state couldn't find enough experienced programmers to upgrade the COBOL programs.

BASIC (Beginner's All-purpose Symbolic Instruction Code), developed at Dartmouth in 1964 by John Kemeny and Tom Kurtz, is another language of the time. BASIC was meant to be an easy language for teaching programming. It was especially simple and required limited computing resources, so it was the first high-level language available on the first personal computers. In fact, Bill Gates and Paul Allen, the founders of Microsoft, got their start by writing a BASIC compiler for the Altair microcomputer in 1975; it was their company's first product. Today one major strain of BASIC is still actively supported as Microsoft Visual Basic.

In the early days when computers were expensive yet slow and limited, there was a concern that programs written in high-level languages would be too inefficient, because compilers could not produce compact and efficient assembly code as well as a skilled assembly language programmer could. Compiler writers worked hard to generate code that was as good as hand-written, which helped to establish the languages. Today, with computers millions of times faster and with plentiful memory, programmers rarely worry about efficiency at the level of individual instructions, though compilers and compiler writers certainly still do.

Fortran, COBOL and BASIC achieved part of their success by focusing on specific application areas and intentionally did not try to handle every possible programming task. During the 1970s, languages were created that were intended for "system programming," that is, for writing programmer tools like assemblers, compilers, text editors, and even operating systems. By far the most successful of these languages was C, developed at Bell Labs by Dennis Ritchie in 1973 and still one of the most popular and widely used. C has changed only a small amount since then; a C program today looks much like one from 30 or 40 years earlier. For comparison, Figure 5.3 shows the same "add up the numbers" program in C.

```
#include <stdio.h>
int main() {
    int num, sum;
    sum = 0;
    while (scanf("%d", &num) != EOF && num != 0)
        sum = sum + num;
    printf("%d\n", sum);
    return 0;
}
```

Figure 5.3: C program to add up numbers.

The 1980s saw the development of languages like C++ (by Bjarne Stroustrup, also at Bell Labs) that were meant to help manage the complexities of very large programs. C++ evolved from C and in most cases a C program is also a valid C++ program, as the one in Figure 5.3 is, though not vice versa. Figure 5.4 shows the example of adding up numbers in C++, one of many ways to write it.

```
#include <iostream>
using namespace std;
int main() {
    int num, sum;
    sum = 0;
    while (cin >> num && num != 0)
        sum = sum + num;
    cout << sum << endl;
    return 0;
}
```

Figure 5.4:   C++ program to add up numbers.

Most of the major programs that we use on our own computers today are written in C or C++. I'm writing this book on Macs, where most software is written in C, C++ and Objective-C (a C dialect). The very first draft was in Word, a C and C++ program; today I edit, format and print with C and C++ programs, and make backup copies on Unix and Linux operating systems (both C programs) while surfing with Firefox, Chrome and Edge (all C++).

During the 1990s, more languages were developed in response to the growth of the Internet and the World Wide Web. Computers continued to get faster processors and bigger memories, and programming speed and convenience became more important than machine efficiency; languages like Java and JavaScript make this tradeoff intentionally.

Java was developed in the early 1990s by James Gosling at Sun Microsystems. Its original target was small embedded systems like home appliances and electronic gadgets, where speed didn't matter much but flexibility did. Java was repurposed to run on web pages, where it did not catch on, but it is widely used by web servers: when you visit a site like Ebay, your computer is running C++ and JavaScript, but Ebay might well be using Java to prepare the page it sends to your browser. Java is also the primary language for writing Android apps. Java is simpler than C++ (though evolving towards similar complexity) but more complicated than C. It's also safer than C since it eliminates some dangerous features and has built-in mechanisms to handle error-prone tasks like managing complicated data structures in memory. For that reason, it's also popular as the first language in programming classes.

Figure 5.5 shows the add-up-the-numbers program in Java. The program is wordier than in the other languages, which is not atypical of Java, but it could be made two or three lines shorter by combining a few computations.

This brings up an important general point about programs and programming. There are always many ways to write a program to do a specific task. In this sense, programming is like literary composition. Concerns like style and effective use of language that matter when writing prose are also important when writing programs and help separate truly great programmers from the merely good ones. Because there are so many ways to express the same computation, it's usually not hard to spot a program that has been copied from another program. This point is made strongly at the beginning of every programming course, yet occasionally students think that changing variable names or the placement of lines will be sufficient to disguise plagiarism. Sorry—it doesn't work.

```java
import java.util.*;
class Addup {
    public static void main (String [] args) {
        Scanner keyboard = new Scanner(System.in);
        int num, sum;
        sum = 0;
        num = keyboard.nextInt();
        while (num != 0) {
            sum = sum + num;
            num = keyboard.nextInt();
        }
        System.out.println(sum);
    }
}
```

Figure 5.5: Java program to add up numbers.

JavaScript is a language in the same broad family that began with C, though with plenty of differences. It was created at Netscape in 1995 by Brendan Eich. Except for sharing part of its name, JavaScript has no relationship to Java. It was designed from the beginning to be used in a browser to achieve dynamic effects on web pages; today almost all web pages include some JavaScript code. We will talk more about JavaScript in Chapter 7, but to make the side-by-side comparisons easy, Figure 5.6 is a version of adding up the numbers in JavaScript.

```javascript
var num, sum;
sum = 0;
num = prompt("Enter new value, or 0 to end");
while (num != '0') {
    sum = sum + parseInt(num);
    num = prompt("Enter new value, or 0 to end");
}
alert(sum);
```

Figure 5.6: JavaScript program to add up numbers.

JavaScript is easy to experiment with. The language itself is simple. You don't need to download a compiler; there's one built into every browser. The results of your computations are immediately visible. As we'll see shortly, you could add a handful of additional lines and put this example on a web page for anyone in the world to use.

Python was created in 1990 by Guido van Rossum at the Centrum Wiskunde & Informatica (CWI) in Amsterdam. It is syntactically somewhat different from C, C++, Java and JavaScript, most visibly in that instead of braces, it uses indentation to indicate how statements are grouped.

Python was designed from the beginning with a focus on readability. It is easy to learn, and has become one of the most widely used of all languages, with a rich collection of software libraries for almost any conceivable programming task. If I had to pick a single language to learn or to teach, I'd pick Python. We will talk more about it in Chapter 7. Meanwhile, Figure 5.7 is a version of adding up the numbers in Python.

```
sum = 0
num = input()
while num != '0':
  sum = sum + int(num)
  num = input()
print(sum)
```

Figure 5.7: Python program to add up numbers.

Where will languages go from here? My guess is that we will continue to make programming easier by using more computer resources to help us. We'll also continue to evolve towards languages that are safer for programmers. For example, C is an exceedingly sharp tool and it's easy to inadvertently make programming errors that aren't detected until too late, perhaps after they have been exploited for nefarious purposes. Newer languages make it easier to prevent or at least detect some errors, though sometimes at the cost of running slower and using more memory. Most of the time this is the right tradeoff to make, though there are certainly still plenty of applications—control systems in cars, planes, spacecraft and weapons, for instance—where tight, fast code matters a lot and highly efficient languages like C will still be used.

Although all languages are formally equivalent, in that they can be used to simulate or be simulated by a Turing machine, they are by no means equally good for all programming tasks. There's a world of difference between writing a JavaScript program to control a complicated web page, and writing a C++ program that implements a JavaScript compiler. It would be unusual to find a programmer who was truly expert at both of these tasks. Experienced professional programmers might be comfortable and passably proficient in a dozen languages, but they are not going to have the same skill in all of them.

Thousands of programming languages have been invented over the years, though fewer than a hundred are in widespread use. Why so many? As I've hinted, each language represents a set of tradeoffs among concerns like efficiency, expressiveness, safety, and complexity. Many languages are explicitly a reaction to the perceived weaknesses of earlier languages, taking advantage of hindsight and more computing power, and often strongly influenced by the personal taste of their designer. New application areas also give rise to new languages that focus on the new domain.

No matter what happens, programming languages are an important and fascinating part of computer science. As the American linguist Benjamin Whorf said, "Language shapes the way we think and determines what we can think about." Linguists still debate whether this is true for natural languages, but it does seem to apply to the artificial languages that we invent to tell our computers what to do.

5.3 Software Development

Programming in the real world tends to happen on a large scale. The strategy is similar to what one might use to write a book or undertake any other big project: figure out what to do, starting with a broad specification that is broken into smaller and smaller pieces, then work on the pieces separately while making sure that they hang together. In programming, pieces tend to be of a size such that one person can write the precise computational steps in some programming language. Ensuring that the pieces written by different programmers work together is challenging, and failing to get this right is a major source of errors. For instance, NASA's Mars Climate Orbiter failed in 1999 because the flight system software used metric units for thrust, but course-correction data was entered in English units, causing an erroneous trajectory that brought the Orbiter too close to the planet's surface.

The examples above that illustrate different languages are mostly less than ten lines long. Small programs of the kind that might be written in an introductory programming course will have a few dozen to a few hundred lines of code. The first "real" program I ever wrote—real in the sense that it was used by a significant number of other people—was about a thousand lines of Fortran. It was a simple-minded word processor for formatting and printing my thesis, and it was taken over by a student agency and used for another five years or so after I graduated. The good old days!

A more substantial program to do a useful task today might have thousands to tens of thousands of lines. Students in my project courses, working in small groups, routinely turn out two or three thousand lines in 8 to 10 weeks, including the time to design their system and learn a new language or two while keeping up with other courses and their extracurricular activities. The product is often a web service for easy access to some university database or a phone app to facilitate social life.

A compiler or a web browser might have hundreds of thousands to a million lines. Big systems, however, have multiple millions or even tens of millions of lines of code, with hundreds or thousands of people working on them at the same time, and lifetimes measured in decades. Companies are usually circumspect about revealing how big their programs are, but reliable information occasionally surfaces. For instance, Google had about two billion lines of code in total, according to a Google conference presentation in 2015; it's likely at least twice that by now.

Software on this scale requires teams of programmers, testers and documenters, with schedules, deadlines, layers of management, and endless meetings to keep it all going. A colleague who was in a position to know used to claim that there was one meeting for every line of code in a major system that he had worked on. Since the system had several million lines, perhaps he exaggerated, but experienced programmers might say "not by much."

5.3.1 Libraries, interfaces, and development kits

If you're going to build a house today, you don't start by cutting down trees to make lumber and digging clay to make your own bricks. Instead, you buy prefabricated pieces like doors, windows, plumbing fixtures, a furnace, and a water heater. House construction is still a big job, but it's manageable because you can build on the work of others and rely on an infrastructure, indeed an entire industry, that will help.

The same is true of programming. Hardly any significant program is created from nothing. Many components written by others can be taken off the shelf and used directly. For instance, if you're writing a program for Windows or a Mac, you have access to code for prefabricated menus, buttons, graphics computations, network connections, database access, and so on. Much of the job is understanding the components and gluing them together in your own way. Of course these components in turn rest on other simpler and more basic ones, often for several layers. Below that, everything runs on the operating system, a program that manages the hardware and controls everything that happens. We'll talk about operating systems in the next chapter.

At the simplest level, programming languages provide a *function* mechanism that makes it possible for one programmer to write code that performs a useful operation, then package it in a form that other programmers can use in their programs without having to know how it works. For example, the C program a few pages ago includes these lines:

```
while (scanf("%d", &num) != EOF && num != 0)
    sum = sum + num;
printf("%d\n", sum);
```

This code "calls" (that is, uses) two functions that come with C: scanf reads data from an input source, which is analogous to GET in the Toy, and printf prints output, like PRINT. A function has a name and a set of input data values that it needs to do its job; it does a computation and may return a result to the part of the program that used it. The syntax and other details here are specific to C and would be different in another language, but the idea is universal. Functions make it possible to create a program by building on components that have been created separately and that can be used as necessary by all programmers.

A collection of related functions is usually called a *library*. For instance, C has a standard library of functions for reading and writing data on disks and other places, and scanf and printf are part of that library.

The services that a function library provides are described to programmers in terms of an *Application Programming Interface* or *API*, which lists the functions, what they do, how to use them in a program, what input data they require, and what values they produce. The API might also describe data structures—the organization of data that is passed back and forth—and various other bits and pieces that all together define what a programmer has to do to request services and what will be computed as a result. This specification must be detailed and precise, since in the end the program will be interpreted by a dumb literal computer, not by a friendly and accommodating human.

An API includes not only the bare statement of syntactic requirements but also supporting documentation to help programmers use the system effectively. Large systems today often involve a *Software Development Kit* or *SDK* so programmers can navigate increasingly complicated software libraries. For example, Apple provides an environment and supporting tools for developers writing iPhone and iPad code; Google provides an analogous SDK for Android phones; Microsoft provides a variety of development environments for writing Windows code in different languages for different devices. SDKs are themselves large software systems; for instance, Android Studio, the development environment for Android, is 1.6 GB, and Xcode, the SDK for Apple developers, is much bigger.

5.3.2 Bugs

Sadly, no substantial program works the first time; life is too complicated and programs reflect that complexity. Programming requires perfect attention to detail, something that few people can achieve. Thus all programs of any size will have errors, that is, they will do the wrong thing or produce the wrong answer under some circumstances. Those flaws are called *bugs*, a term popularly attributed to Grace Hopper, who was mentioned above. In 1947, Hopper's colleagues found a literal bug (a dead moth) in the Harvard Mark II, a mechanical computer that they were working with, and she apparently said that they were "debugging" the machine. The bug was preserved and has passed into a kind of immortality; it can be seen in the Smithsonian's American History museum in Washington and in the photograph in Figure 5.8.

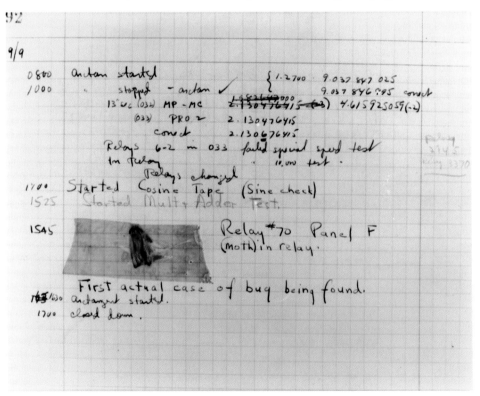

Figure 5.8: Bug from the Harvard Mark II.

Hopper did not coin this use of the word "bug," however; it dates from 1889. As the Oxford English Dictionary (second edition) says,

> **bug.** A defect or fault in a machine, plan, or the like. orig. U.S.
> 1889 Pall Mall Gaz. 11 Mar. 1/1 Mr. Edison, I was informed, had been up the two previous nights discovering 'a bug' in his phonograph—an expression for solving a difficulty, and implying that some imaginary insect has secreted itself inside and is causing all the trouble.

Bugs arise in so many ways that it would take a large book to describe them (and such books do exist). Among the endless possibilities are forgetting to handle a case that might occur, writing the wrong logical or arithmetic test to evaluate some condition, using the wrong formula, accessing memory outside the range allotted to the program or a part of it, applying the wrong operation to a particular kind of data, and failing to validate user input.

As a contrived example, Figure 5.9 shows a pair of JavaScript functions that convert from Celsius temperatures to Fahrenheit and vice versa. (The operators * and / perform multiplication and division respectively.) One of these functions has an error. Can you see it? I'll come back to it in a moment.

```
function ctof(c) {
    return 9/5 * c + 32;
}
function ftoc(f) {
    return 5/9 * f - 32;
}
```

Figure 5.9: Functions for converting between Celsius and Fahrenheit.

Testing is a big part of real-world programming. Software companies often have more tests than code and more testers than programmers, in the hope of identifying as many bugs as possible before software is shipped to users. That's hard, but one can at least get to the state where bugs are encountered infrequently.

How would you test the temperature conversion functions in Figure 5.9? You would certainly want to try some simple test cases with known answers, for example Celsius temperatures of 0 and 100 where the corresponding Fahrenheit values should be 32 and 212. Those work fine.

But in the other direction, Fahrenheit to Celsius, things don't work so well: the function reports that 32°F is −14.2°C and 212°F is 85.8°C, both grossly wrong. The problem is that parentheses are necessary to subtract 32 from the Fahrenheit value before multiplying it by 5/9; the expression in `ftoc` should read

```
    return 5/9 * (f - 32);
```

Fortunately these are easy functions to test, but you can imagine how much work might be involved in testing and debugging a program with a million lines when the failures are not so obvious.

By the way, these two functions are inverses of each other (like 2^n and $\log n$) and that makes some testing easy. If you pass any value through each function in turn, the result should be the original number, except perhaps for a tiny discrepancy caused by the fact that computers don't represent non-integers with perfect precision.

Bugs in software can leave systems vulnerable to attack, often by permitting adversaries to overwrite memory with their own malicious code. There's an active marketplace in exploitable bugs; white hats fix problems, black hats exploit them, and there's a gray area in the middle where government agencies like the NSA stockpile exploits to be used or fixed later.

The prevalence of vulnerabilities explains the frequent updates of important programs like browsers, which are the focus of attention for many hackers. Writing robust programs is hard, and bad guys are always watching for openings; it's important for ordinary users to keep our software up to date as security holes are patched.

Another complexity in real-world software is that the environment changes all the time, and programs have to be adapted. New hardware is developed; it needs new software that may require changes in systems. New laws and other requirements change the specifications of programs—for instance, a program like TurboTax has to respond to frequent tax law changes in many jurisdictions. Computers, tools, languages and physical devices become obsolete and have to be replaced. Data formats become obsolete too—for example, Word files from the early 1990s can't be read by today's version of Word. Expertise disappears too, as people retire, die, or get fired in a corporate downsizing. Student-created systems at universities suffer in the same way when the expertise graduates.

Keeping up with steady change is a big part of software maintenance, but it has to be done; otherwise, programs suffer "bit rot" and after a while don't work any more or can't be updated because they can't be recompiled or some library has changed too much. At the same time, the effort to fix problems or add new features can create new bugs or change behaviors that users depended on.

5.4 Intellectual Property

The term *intellectual property* refers to various kinds of intangible property that result from individual creative efforts like inventions or authorship—books, music, paintings, photographs. Software is an important example. It's intangible, yet valuable. It takes sustained hard work to create and maintain a significant body of code. At the same time, software can be copied in unlimited quantities and distributed worldwide at zero cost, it's readily modified, and ultimately it's invisible.

Ownership of software raises difficult legal issues, I think more so than hardware does, though that may be my bias as a programmer. Software is a newer field than hardware; there was no software before about 1950, and it's only in the past roughly 40 years that software has become a major independent economic sector. As a result, there has been less time for laws, commercial practice and social norms to evolve. In this section, I'll discuss some of the problems, to give you enough technical background that you can at least appreciate the situation from multiple viewpoints. I'm also writing from the perspective of US law; other countries have analogous systems but differ in many respects.

Several legal mechanisms for protecting intellectual property apply, with varying degrees of success, to software. These include trade secrets, trademarks, copyrights, patents, and licenses.

5.4.1 Trade secret

Trade secret is the most obvious. The property is kept secret by its owner, or disclosed to others only under a legally binding contract like a non-disclosure agreement. This is simple and often effective, but provides little recourse if the secret is ever revealed. The classic example of a trade secret, in a different domain, is the formula for Coca-Cola. In theory, if the secret became public knowledge, anyone could make an identical product, though they could not call it Coca-Cola or Coke, because those are trademarks, another form of intellectual property. In software, the code for major systems like PowerPoint or Photoshop is a trade secret.

5.4.2 Trademark

A *trademark* is a word or phrase, a name, a logo, even a distinctive color, that distinguishes the goods or services provided by a company. For example, think of the flowing script in which the words Coca-Cola appear in advertisements, and the shape of the classic Coke bottle; both are trademarks. McDonald's Golden Arches are a trademark that distinguishes them from other fast-food companies.

Computing has endless trademarks, like the glowing cutout on Mac laptops, a trademark of Apple. Microsoft's four-color logos on their operating systems, computers and game controllers are also examples.

5.4.3 Copyright

Copyright protects creative expression. Copyright is familiar in the context of literature, art, music and movies—it protects creative work from being copied by others, at least in theory, and it gives creators the right to exploit their work for a limited period. In the US, that period used to be 28 years with one renewal but is now the lifetime of the author plus 70 years. In many other countries, the period is life plus 50 years. In 2003, the US Supreme Court ruled that 70 years after the author's death is a "limited" term. This is technically correct, but practically not very different from "forever." Rights-holders in the US are pushing hard to extend copyright terms worldwide to conform to US law.

Enforcing copyright for digital material is difficult. Any number of electronic copies can be made and distributed throughout the online world at no cost. Attempts to protect copyrighted material by encryption and other forms of *digital rights management* or *DRM* have uniformly failed—the encryption usually proves breakable and even if not, material can be re-recorded as it is being played (the "analog hole"), for example by surreptitiously filming in a theater. Legal recourse against copyright infringement is hard for individuals and even for large organizations to pursue effectively. I'll return to this topic in Chapter 9.

Copyright also applies to programs. If I write a program, I own it, just as if I had written a novel. No one else can use my copyrighted program without my permission. That sounds simple enough, but as always the devil is in the details. If you study my program's behavior and create your own version, how similar can it be without violating my copyright? If you change the formatting and the names of all the variables in the program, that's still a violation. However, for more subtle changes it's not obvious and the issues can only be settled by an expensive legal process. If you study the behavior of my program, understand the behavior thoroughly, and then do a genuinely new implementation, that might be valid. In fact, in the technique called *cleanroom development* (a reference to integrated-circuit manufacturing), programmers explicitly have no access to or knowledge of the code whose properties they are trying to replicate. They write new code that behaves the same way as the original but which has demonstrably not been copied. Then the legal question becomes one of proving that the cleanroom really was clean and no one was tainted by exposure to the original code.

5.4.4 Patent

Patents provide legal protection for inventions. This contrasts with copyrights, which protect only expression—how the code is written—but not any original ideas the code might contain. There are plenty of hardware patents like the cotton gin, the telephone, the transistor, the laser, and of course myriad processes, devices, and improvements on them.

Originally, software—algorithms and programs—was not patentable, since it was thought to be "mathematics" and thus not within the scope of patent law. As a programmer with a modest background in mathematics, I don't think that algorithms are mathematics, though they might involve mathematics. (Think about Quicksort, which today might well be patentable.) Another viewpoint is that many software patents are obvious, no more than using a computer to do some straightforward or well-known process, and thus should not be patentable because they lack originality. I'm much more in sympathy with that position, though again as a non-expert and certainly not as a lawyer.

The poster child for software patents might be Amazon's "1-click" patent. In September 1999, US patent 5,960,411 was granted to four inventors at Amazon.com, including Jeff Bezos, the founder and CEO. The patent covers "A method and system for placing an order to purchase an item via the Internet"; the claimed innovation was allowing a registered customer to place an order with a single mouse click (Figure 5.10). By the way, note that "1-Click" is a registered Amazon trademark, indicated by 1-Click®.

Figure 5.10: Amazon 1-Click®.

The 1-click patent was the subject of debate and legal struggle for nearly 20 years. It's probably fair to say that most programmers think the idea is obvious, but the law requires that an invention be "unobvious" to "a person having ordinary skill in the art" at the time of invention, which was 1997, in the early days of web commerce. The US Patent Office denied some claims of the patent; others were sustained on appeal. In the meantime, the patent was licensed by other companies, including Apple for its iTunes online store, and Amazon obtained injunctions against companies that used the 1-click idea without permission. Naturally the situation was different in other countries. Fortunately, this is all moot today since the duration of a patent is 20 years, and it has now expired.

One of the downsides of how easy it is to get a software patent has been the rise of so-called *patent trolls* or, less pejoratively, "non-practicing entities." A patent troll acquires the rights to a patent, not to use the invention but to sue others that it claims are in violation. The suit is often filed in a location where judgments have tended to favor the plaintiff, that is, the troll. The direct cost of patent litigation is high and the cost if one loses a suit is potentially very high. Especially for small companies it's easier and safer to cave in and pay a license fee to the troll, even though the patent claims are weak and the infringement is not clear.

The legal climate is changing, though slowly, and this kind of patent activity may become less of an issue, but it is still a major problem.

5.4.5 Licenses

Licenses are legal agreements granting permission to use a product. Every computer user is familiar with one step in the process of installing a new version of some software: the "End User License Agreement" or EULA. A dialog box shows a small window on an enormous block of tiny type, a legal document whose terms you must agree to before you can go further. Most people just click to get past it and thus are in principle and probably in practice legally bound by the terms of the agreement.

If you do read those terms, it won't be a big surprise to discover that they are one-sided. The supplier disclaims all warranties and liability, and in fact doesn't even promise that the software will do anything. The excerpt below (all capital letters as in the original) is a small part of the EULA for macOS Mojave, the operating system running on my Mac:

B. YOU EXPRESSLY ACKNOWLEDGE AND AGREE THAT, TO THE EXTENT PERMITTED BY APPLICABLE LAW, USE OF THE APPLE SOFTWARE AND ANY SERVICES PERFORMED BY OR ACCESSED THROUGH THE APPLE SOFTWARE IS AT YOUR SOLE RISK AND THAT THE ENTIRE RISK AS TO SATISFACTORY QUALITY, PERFORMANCE, ACCURACY AND EFFORT IS WITH YOU.

C. TO THE MAXIMUM EXTENT PERMITTED BY APPLICABLE LAW, THE APPLE SOFTWARE AND SERVICES ARE PROVIDED "AS IS" AND "AS AVAILABLE", WITH ALL FAULTS AND WITH-OUT WARRANTY OF ANY KIND, AND APPLE AND APPLE'S LICENSORS (COLLECTIVELY REFERRED TO AS "APPLE" FOR THE PURPOSES OF SECTIONS 7 AND 8) HEREBY DISCLAIM ALL WARRANTIES AND CONDITIONS WITH RESPECT TO THE APPLE SOFTWARE AND SER-VICES, EITHER EXPRESS, IMPLIED OR STATUTORY, INCLUDING, BUT NOT LIMITED TO, THE IMPLIED WARRANTIES AND/OR CONDITIONS OF MERCHANTABILITY, SATISFACTORY QUAL-ITY, FITNESS FOR A PARTICULAR PURPOSE, ACCURACY, QUIET ENJOYMENT, AND NON-INFRINGEMENT OF THIRD PARTY RIGHTS.

D. APPLE DOES NOT WARRANT AGAINST INTERFERENCE WITH YOUR ENJOYMENT OF THE APPLE SOFTWARE AND SERVICES, THAT THE FUNCTIONS CONTAINED IN, OR SERVICES PER-FORMED OR PROVIDED BY, THE APPLE SOFTWARE WILL MEET YOUR REQUIREMENTS, THAT THE OPERATION OF THE APPLE SOFTWARE OR SERVICES WILL BE UNINTERRUPTED OR ERROR-FREE, THAT ANY SERVICES WILL CONTINUE TO BE MADE AVAILABLE, THAT THE APPLE SOFTWARE OR SERVICES WILL BE COMPATIBLE OR WORK WITH ANY THIRD PARTY SOFTWARE, APPLICATIONS OR THIRD PARTY SERVICES, OR THAT DEFECTS IN THE APPLE SOFTWARE OR SERVICES WILL BE CORRECTED. INSTALLATION OF THIS APPLE SOFTWARE MAY AFFECT THE AVAILABILITY AND USABILITY OF THIRD PARTY SOFTWARE, APPLICA-TIONS OR THIRD PARTY SERVICES, AS WELL AS APPLE PRODUCTS AND SERVICES.

Most EULAs say you can't sue for damages if the software does you harm. There are conditions on what the software can be used for, and you agree that you won't try to reverse-engineer or disassemble it. You can't ship it to certain countries and you can't use it to develop nuclear weapons (really). My lawyer friends say that such licenses are generally valid and enforceable if the terms are not too unreasonable, which seems to beg the question of what is reasonable.

Another clause may come as a bit of a surprise, especially if you have purchased the software in a physical or online store: "This software is licensed, not sold." For most purchases, a legal doctrine called "first sale" says that once you have bought something, you own it. If you buy a printed book, it's your copy and you can give it away or resell it to someone else, though of course you can't violate the author's copyright by making and distributing copies. But suppliers of digital goods almost always "sell" them under a license that lets the supplier retain ownership and restrict what you can do with "your" copy.

A great example of this surfaced in July 2009. Amazon "sells" lots of books for its Kindle e-book readers, but in fact the books are licensed, not sold. At one point, Amazon realized that it was distributing some books that it did not have permission for, so it "unsold" them by disabling them on all Kindles. In a wonderful bit of irony, one of the recalled books was an edition of George Orwell's dystopian novel *1984*. I'm sure that Orwell would have loved the Kindle story.

APIs also raise some interesting legal questions, mostly focused on copyrights. Suppose I'm the manufacturer of a programmable game system, analogous to the Xbox or PlayStation. I want people to buy my game machine, and that will be more likely if there are plenty of good games for it. I can't possibly write all that software myself, so I carefully define a suitable API—an application programming interface— so programmers can write games for my machine. I might also provide a software development kit or SDK, analogous to Microsoft's XDK for the Xbox, to help game developers. With luck, I will sell a bunch of machines, make a pile of money, and retire happy.

An API is in effect a contract between the service user and the service provider. It defines what happens on both sides of the interface—not the details of how it's implemented, but definitely what each function does when used in a program. That means that someone else, like a competitor, could also play the provider side, by building a competing game machine that provides the same API as mine. If they used cleanroom techniques, that would ensure that they didn't copy my implementation in any way. If they did this well—everything works the same—and if the competitor's machine were better in other ways, like a lower price and sexier physical design, it could drive me out of business. That's bad news for my hopes of becoming wealthy.

What are my legal rights? I can't patent the API, since it's not an original idea, and it's not a trade secret because I have to show it to people so they can use it. If defining the API is a creative act, however, I might be able to protect it by copyright, allowing others to use it only if they have licensed the rights from me; the same is likely to be the case if I provide an SDK. Is that sufficient protection? This legal question, and a variety of others like it, is not really resolved.

The copyright status of APIs is not a hypothetical question. In January 2010, Oracle bought Sun Microsystems, creator of the Java programming language, and in August 2010 sued Google, alleging that Google was illegally using the Java API on Android phones, which run Java code.

To over-simplify the tale of a complicated case, a district court determined that APIs were not copyrightable. Oracle appealed and the decision was reversed. Google petitioned the US Supreme Court to hear the case, but in June 2015 the court declined to do so. In the next round, Oracle asked for over 9 billion dollars in damages, but a jury decided that Google's use of the APIs was "fair use" and thus not a violation of copyright law. I think that most programmers would agree with Google in this specific case, but the matter is not yet settled. (As a disclaimer, I have twice been a signatory on amicus briefs submitted by the Electronic Frontier Foundation that support Google's position.) After yet more rounds of legal process, the Supreme Court heard the case again in October, 2020.

5.5 Standards

A *standard* is a precise and detailed description of how some artifact is built or is supposed to work. Some standards, like the Word `.doc` and `.docx` file formats, are *de facto* standards—they have no official standing but everyone uses them. The word "standard" is best reserved for formal descriptions, often developed and maintained by a quasi-neutral party like a government agency or a consortium, that define how something is built or operates. The definition is sufficiently complete and precise that separate entities can interact or provide independent implementations.

We benefit from hardware standards all the time, though we may not notice how many there are. If I buy a new television set, I can plug it into the electrical outlets in my home, thanks to standards for the size and shape of plugs and the voltage they provide. (Though not in other countries, of course; for European vacations, I have to take along several ingenious adapters that let me plug my North American power supplies into the different sockets in England and France.) The TV itself will receive signals and display pictures because of standards for broadcast and cable television. I can plug other devices into it through standard cables and connectors like HDMI, USB, S-video and so on. But every TV needs its own remote control because those are not standardized; so-called "universal" remotes work only some of the time.

There are sometimes even competing standards, which seems counter-productive. (As computer scientist Andy Tanenbaum once said, "The nice thing about standards is that you have so many to choose from.") Historical examples include Betamax versus VHS for videotape and HD-DVD versus Blu-ray for high-definition video disks. In both cases, one standard eventually won out, but in other cases multiple standards may co-exist, like the two incompatible cell phone technologies used in the US until around 2020.

Software has plenty of standards as well, including character sets like ASCII and Unicode, programming languages like C and C++, algorithms for encryption and compression, and protocols for exchanging information over networks.

Standards are crucial for interoperability and an open competitive landscape. They make it possible for independently created things to cooperate, and they open an area to competition from multiple suppliers, while proprietary systems tend to lock everyone in. Naturally the owners of proprietary systems prefer lock-in. Standards have disadvantages too—a standard can impede progress if it is inferior or outdated yet everyone is forced to use it. But these are modest drawbacks compared to the advantages.

5.6 Open Source Software

The code that a programmer writes, whether in assembly language or (much more likely) in a high-level language, is called *source code*. The result of compiling it into a form suitable for a processor to execute is called *object code*. This distinction, like several others I've made, might seem pedantic but it's important. Source code is readable by programmers, though perhaps with some effort, so it can be studied and adapted, and any innovations or ideas it contains are visible. By contrast, object code has gone through so much transformation that it is usually impossible to recover anything remotely like the original source code or to extract any form that can be used for making variants or even understanding how it works. It is for this reason that most commercial software is distributed only in object-code form; the source code is a valuable secret and is kept metaphorically and perhaps literally under lock and key.

Open source refers to an alternative in which source code is freely available for study and improvement.

In early times, most software was developed by companies and most source code was unavailable, a trade secret of whoever developed it. Richard Stallman, a programmer working at MIT, was frustrated that he could not fix or enhance the programs that he was using because their source code was proprietary and thus inaccessible to him. In 1983, Stallman started a project that he called *GNU* ("GNU's Not Unix," gnu.org) to create free and open versions of important software systems, like an operating system and compilers for programming languages. He also formed a non-profit organization called the Free Software Foundation to support open source. The goal was to produce software that was perpetually "free" in the sense of being non-proprietary and unencumbered by restrictive ownership. This was accomplished by distributing implementations under a clever copyright license called the GNU General Public License or GPL.

The Preamble to the GPL says

> "The licenses for most software and other practical works are designed to take away your freedom to share and change the works. By contrast, the GNU General Public License is intended to guarantee your freedom to share and change all versions of a program—to make sure it remains free software for all its users."

The GPL specifies that the licensed software can be freely used, but if it is distributed to anyone else the distribution must make the source code available with the same "free for any use" license. The GPL is strong enough that companies that have violated its terms have been forced by court decisions to stop their use of the code or to distribute source that they based on licensed code.

The GNU project, supported by companies, organizations, and individuals, has produced a large collection of program development tools and applications, all covered by the GPL. Other open source programs and documents have analogous licenses, for example the Creative Commons that accompanies many images in Wikipedia. In some cases, the open source versions set the standard against which proprietary commercial versions are measured. The Firefox and Chrome browsers are open source; so are Apache and NGINX, the two most common web servers; so is the Android operating system for cell phones.

Programming languages and supporting tools are now almost always open source; indeed, it would be hard to establish a new programming language if it were strictly proprietary. In the past decade, Google created and released Go, Apple created and released Swift, Mozilla created and released Rust, and Microsoft released C# and F#, which had been proprietary for years.

The Linux operating system is perhaps the most visible open source project; it is widely used by individuals and large commercial enterprises like Google, which runs its entire infrastructure on Linux. You can download the Linux operating system source code for free from `kernel.org`; you can use it for your own purposes and make any modifications you like. But if you distribute it in any form, for example, in a new gadget that has an operating system, you must make your source code available under the same GPL. Both my cars, from different manufacturers, run Linux; deep within the on-screen menu system is a GPL statement and a link. Using that link, I was able to download the code from the Internet (not from the car!), nearly 1 GB of Linux source.

Open source is intriguing. How can one make money by giving software away? Why do programmers voluntarily contribute to open source projects? Can open source written by volunteers be better than proprietary software developed by large teams of coordinated professionals? Is the availability of source code a threat to national security?

These questions continue to interest economists and sociologists, but some answers are becoming clearer. For example, Red Hat was founded in 1993 and by 1999 was a public company traded on the New York Stock Exchange; in 2019 it was acquired by IBM for $34 billion. Red Hat distributes Linux source code that you can get for free on the Internet but makes money by charging for support, training, quality assurance, integration, and other services. Many open source programmers are regular employees of companies that use open source and contribute to it. IBM, Facebook, and Google are notable examples but certainly not unique; Microsoft is now one of the largest contributors to open source software projects. The companies benefit from being able to help guide the evolution of programs and from having others fix bugs and make enhancements.

Not all open source software is best of breed, and the open source versions of some software may lag behind the commercial systems on which they are modeled. Nevertheless, for core programmer tools and systems, open source is hard to beat.

5.7 Summary

Programming languages are how we tell our computers what to do. Although the idea can be pushed too far, there are parallels between natural languages and the artificial languages that we invent to make it easier to write code. One obvious parallel is that there are thousands of programming languages, although probably no more than a few hundred programming languages are in frequent use, and two dozen account for the large majority of programs that run today. Of course programmers hold opinions, often strong, about which languages are best, but one of the reasons why there are so many languages is that no one language is ideal for all programming tasks. There is always the feeling that a suitable new language would make programming ever so much easier and more productive. Languages have also evolved to take advantage of steadily increasing hardware resources. Long ago, programmers had to work hard to squeeze programs into available memory; that's less of an issue today, and languages provide mechanisms that manage memory use automatically so that programmers don't have to think about it as much.

Intellectual property issues for software are challenging, especially with patents, where trolls are a strongly negative force. Copyright seems easier, but even there, major legal issues like the status of APIs remain unresolved. As is often the case, the law does not (and probably can not) respond quickly to new technology, and when the responses come, they vary from country to country.

6

Software Systems

"The programmer, like the poet, works only slightly removed from pure thought-stuff. He builds his castles in the air, from air, creating by exertion of the imagination. Few media of creation are so flexible, so easy to polish and rework, so readily capable of realizing grand conceptual structures."

Frederick P. Brooks, *The Mythical Man-Month*, 1975.

In this chapter, we're going to look at two major kinds of software: operating systems and applications. As we'll see, an *operating system* is the software underpinning that manages the hardware of a computer and makes it possible to run other programs, which are called *applications*.

When you use a computer at home, school or office, you have a wide variety of programs available, including browsers, word processors, music and movie players, tax software (alas), virus scanners, plenty of games, and tools for mundane tasks like searching for files or looking at folders. The situation is analogous on your phone, though with different details.

The jargon term for such programs is *application*, presumably from "this program is an application of a computer to some task." It's a standard term for programs that are more or less self-contained and focused on doing a single job. The word used to be the province of computer programmers, but with the runaway success of Apple's App Store, which sells applications to run on iPhones, the abbreviated form *app* has become part of everyone's vocabulary.

When you buy a new computer or a phone, it comes with a number of such programs already installed, and more are added over time as you buy them or download them. Apps in this sense are important to us as users, and they have interesting properties from several technical perspectives. We're going to talk briefly about a few apps, then focus on a specific one, the browser. A browser is a representative example that everyone is familiar with, but it still holds some surprises, including an unexpected parallel with operating systems.

Let's begin, however, with the behind-the-scenes program that makes it possible to use applications: the operating system. As we go, keep in mind that pretty much every computer, whether laptop, phone, tablet, media player, smart watch, camera or other gadget, has an operating system of some type to manage the hardware.

6.1 Operating Systems

In the early 1950s, there was no distinction between application and operating system. Computers were so limited that they could only run one program at a time, and that program took over the whole machine; indeed, programmers had to sign up for a time slot to use the computer (in the middle of the night if one were a lowly student) to run their one program. As computers became more sophisticated, it was too inefficient to have amateurs running them, so the job was turned over to professional operators, who fed programs in and distributed the results. Operating systems began as programs to help automate those tasks for the human operators.

Operating systems steadily became more elaborate, matching the evolution of the hardware that they controlled, and as hardware became more powerful and complicated, it made sense to devote more resources to controlling it. The first widely used operating systems appeared in the late 1950s and early 1960s, usually provided by the same company that made the hardware and tightly tied to it by being written in assembly language. Thus IBM and smaller companies like Digital Equipment and Data General provided their own operating systems for their own hardware. Fred Brooks, who is quoted in the epigraph above, managed the development of IBM's System/360 series of computers and OS/360, the company's flagship operating system from 1965 to 1978. Brooks won the 1999 Turing Award for his contributions to computer architecture, operating systems, and software engineering.

Operating systems were also objects of research at universities and industrial labs. MIT was a pioneer, creating in 1961 a system called CTSS ("Compatible Time-Sharing System") that was especially advanced for its time, and, unlike its industrial competitors, a pleasure to use. The Unix operating system was created at Bell Labs starting in 1969 by Ken Thompson and Dennis Ritchie, who had worked on Multics, the more elaborate but less successful follow-on to CTSS. Today most operating systems, except those from Microsoft, are descended either from the original Bell Labs Unix or the compatible but independently created Linux. Ritchie and Thompson shared the 1983 Turing Award for their creation of Unix.

A modern computer is a complicated beast indeed. It has many parts—processors, memory, secondary storage, display, network interfaces, and on and on—as we saw in Figure 1.2. To use these components effectively, it needs to run multiple programs at the same time, some of which are waiting for something to happen (a web page to download), some of which are demanding an immediate response (tracking mouse movement or updating the display as you play a game), and some of which are interfering with the others (starting up a new program that needs space in the already over-crowded memory). It's a mess.

The only way to manage this elaborate juggling act is to use a program, another example of making the computer help with its own operation. That program is called an operating system. For computers at home or work, Microsoft Windows, in its various evolutionary stages, is the most common operating system; it runs perhaps 80 to 90 percent of the desktop and laptop computers one sees in day-to-day life. Apple computers run macOS. Many behind-the-scenes computers (and some foreground ones as well) run Linux. Cell phones run operating systems too, originally specialized systems but today often smaller versions of Unix or Linux. For instance, iPhones and iPads run iOS, an operating system derived from macOS, which at its heart is a Unix variant, while Android phones run Linux, as does my television, the TiVo, Amazon's Kindle, and Google Nest. I can even log in to my Android phone and run basic Unix commands on it.

An operating system controls and allocates the resources of a computer. First, it manages the processor, scheduling and coordinating the programs that are currently in use. It switches the processor's attention among the programs that are actively computing at any given moment, both applications and background processes like anti-virus software. It suspends programs that are waiting for an event, like a user clicking on a dialog box. It keeps individual programs from hogging resources—if one program demands too much processor time, the operating system throttles it back so other tasks get a reasonable share as well.

A typical operating system will have hundreds of processes in simultaneous operation. Some are programs started by users, though most are system tasks that are invisible to the casual user. You can see what's going on with programs like Activity Monitor on macOS and Task Manager on Windows, or similar programs on your phone. Figure 6.1 shows a handful of the 300 processes running on the Mac where I am currently typing. Most of these are independent of each other and thus a good match for a multi-core architecture.

Second, the operating system manages primary memory. It loads programs into memory so they can begin executing instructions. It copies them out to the disk temporarily if there isn't enough memory for everything that's happening at the same time, then moves them back in when there's room again. It keeps separate programs from interfering with each other so one program can't access the memory allocated to another program or to the operating system itself. This is partly to maintain sanity but it's also a safety measure: one doesn't want a rogue or buggy program poking around where it shouldn't be. (The "blue screen of death" that used to be a common sight in Windows was sometimes caused by a failure to provide adequate protection.)

It takes good engineering to make effective use of primary memory. One technique is to bring only part of a program into memory when needed and copy it back out to disk when inactive, a process called *swapping*. Programs are written as if they had the entire computer to themselves and unlimited primary memory. A combination of software and hardware provides this abstraction, which makes programming significantly easier. The operating system then has to support the illusion by swapping chunks of program in and out, with help from the hardware to translate program memory addresses into real addresses in real memory. This mechanism is called *virtual memory*. Like most uses of the word "virtual," it means giving the illusion of reality but not the real thing.

Figure 6.2 shows how my computer is using its memory. Processes are sorted by the amount of memory they are using. In this case, browser processes account for most of the memory use, which is typical—browsers are memory-hungry. As a general rule, the more memory you have, the faster your computer will feel, since it will be spending less time swapping between memory and secondary storage. If you

Process Name	% CPU ⌄	Real Mem	Rcvd Bytes	Sent Bytes	CPU Time	Threads
WindowServer	18.8	72.2 MB	0 bytes	0 bytes	1:34:44.28	10
Activity Monitor	10.6	192.0 MB	0 bytes	0 bytes	3:39:50.44	5
hidd	6.8	6.0 MB	0 bytes	0 bytes	9:28.23	6
kernel_task	5.0	2.53 GB	4.7 MB	951 KB	4:58:20.23	225
AppleUserHIDDrivers	3.1	2.2 MB	0 bytes	0 bytes	21.84	3
Dock	2.2	15.8 MB	0 bytes	0 bytes	34.92	5
Microsoft Edge Helper (R...	1.9	99.1 MB	0 bytes	0 bytes	13:29.42	24
Firefox	1.6	1.04 GB	25 KB	10 KB	36:40.77	68
FirefoxCP WebExtensions	1.5	381.4 MB	0 bytes	0 bytes	18:37.99	37
launchservicesd	1.3	5.4 MB	0 bytes	0 bytes	4:26.46	7
sysmond	1.2	4.9 MB	0 bytes	0 bytes	3:22:03.58	3
Microsoft Edge	1.2	89.5 MB	5 KB	3 KB	15:57.75	76
corespotlightd	0.6	10.6 MB	0 bytes	0 bytes	37.78	5
tccd	0.4	6.1 MB	0 bytes	0 bytes	14.34	3
launchd	0.4	14.2 MB	0 bytes	0 bytes	1:22:21.27	4
loginwindow	0.3	27.3 MB	0 bytes	0 bytes	2:15.00	5

System:	3.34%	CPU LOAD	Threads:	1,955
User:	5.28%		Processes:	414
Idle:	91.38%			

Figure 6.1: Activity Monitor showing processor activity on macOS.

want your computer to run faster, more primary memory is likely to be the most cost-effective thing to try, though there is usually a physical upper limit on how much can be added, and some computers can't be upgraded.

Third, the operating system manages information stored on secondary storage. A major component of the operating system called the *file system* provides the familiar hierarchy of folders and files that we see when we use a computer. We'll come back to file systems later in this chapter, since they have enough interesting properties to warrant a more extended discussion.

Finally, the operating system manages and coordinates the activities of the devices connected to the computer. A program can assume that it has non-overlapping windows all to itself. The operating system performs the complicated task of managing multiple windows on the display, making sure that the right information gets to the right window, and that it's properly restored when the window is moved, resized, or hidden and re-exposed. The operating system directs input from the keyboard and mouse to the program that's expecting it. It handles traffic to and from network connections, whether wired or wireless. It sends data to printers and it fetches data from scanners.

Notice that I said that an operating system is a program. It's just another program, like the ones in the previous chapter, written in the same kinds of languages, most often C or C++. Early operating systems were small, since memories were smaller and the job was simpler. The earliest operating systems only ran one program at a time, so only limited swapping was needed. There wasn't a lot of memory to

Process Name	Memory	Real Me... ⌄	Private Mem	Shared Mem	VM Compressed
kernel_task	144.2 MB	2.53 GB	0 bytes	0 bytes	0 bytes
Firefox	1.16 GB	1.04 GB	903.7 MB	318.4 MB	123.5 MB
FirefoxCP WebExtensions	372.2 MB	398.2 MB	208.6 MB	207.9 MB	101.1 MB
Activity Monitor	161.2 MB	237.3 MB	96.5 MB	34.3 MB	13.1 MB
Microsoft Edge Helper (Renderer)	183.7 MB	169.8 MB	37.7 MB	187.8 MB	89.1 MB
CrashPlanService	621.9 MB	108.5 MB	123.9 MB	4.0 MB	519.5 MB
Preview	88.3 MB	106.4 MB	44.2 MB	62.3 MB	30.7 MB
FirefoxCP Web Content	140.4 MB	106.4 MB	13.1 MB	286.4 MB	45.7 MB
Microsoft Edge Helper (Renderer)	78.5 MB	99.1 MB	23.8 MB	189.2 MB	33.9 MB
Terminal	112.8 MB	97.8 MB	32.0 MB	99.9 MB	25.1 MB
Microsoft Edge	159.9 MB	89.5 MB	56.5 MB	221.7 MB	87.2 MB
softwareupdated	49.4 MB	83.2 MB	35.3 MB	11.2 MB	11.0 MB
WindowServer	540.2 MB	73.0 MB	31.7 MB	138.5 MB	73.0 MB
Safari	38.7 MB	69.1 MB	33.5 MB	25.3 MB	0 bytes
FirefoxCP Web Content	23.9 MB	56.4 MB	22.0 MB	205.6 MB	0 bytes
syspolicyd	51.0 MB	40.3 MB	30.9 MB	4.3 MB	20.0 MB

Activity Monitor (All Processes) — CPU | Memory | Energy | Disk | Network

MEMORY PRESSURE

Physical Memory:	16.00 GB
Memory Used:	7.82 GB
Cached Files:	6.77 GB
Swap Used:	725.8 MB

App Memory:	3.85 GB
Wired Memory:	2.81 GB
Compressed:	1.15 GB

Figure 6.2:   Activity Monitor showing memory use on macOS.

allocate, less than a hundred kilobytes. They didn't have many external devices to deal with, certainly not the rich variety that we have today. Operating systems are now very large—millions of lines of code—and complicated because they are doing a variety of complicated tasks.

For calibration, the 6th edition of the Unix operating system, the ancestor of many systems today, was 9,000 lines of C and assembly language in 1975, and was written by two people. Today Linux has well over 10 million lines, the work of thousands of people over decades. Windows 10 is guessed to be about 50 million lines, though no authoritative size has been published. These numbers aren't directly comparable anyway, since modern computers are much more sophisticated and deal with much more complex environments and far more devices; there are also differences in what's deemed to be included in the operating system.

Since an operating system is just a program, you can in principle write your own. Indeed, Linux began when Linus Torvalds, a Finnish college student, decided to write his own version of Unix from scratch in 1991. He posted an early draft (just under 10,000 lines) on the Internet and invited others to try it and help out. Since then, Linux has become a major force in the software industry, used by many large companies and numerous smaller players. As noted in the previous chapter, Linux is open source, so anyone can use it and contribute. Today there are thousands of contributors, with a core of full-time developers; Torvalds still maintains overall control and is the ultimate arbiter of technical decisions.

You can run a different operating system on your hardware than might have been originally intended—running Linux on computers originally intended for Windows is a good example. You can store several operating systems on disk and determine which one to run each time you turn the computer on. This "multiple boot" feature can be seen with Apple's Boot Camp, which makes it possible to start up a Mac running Windows instead of macOS.

You can even run one operating system under the control of another, as a *virtual operating system*. Virtual operating system programs like VMware, VirtualBox and Xen (which is open source) make it possible to run one operating system, say Windows or Linux, as a guest operating system on a host, say macOS. The host operating system intercepts requests made by the guest that would require operating system privileges, such as file system or network access. The host does the operation and then returns to the guest. When the host and guest are both compiled for the same hardware, the guest system runs at the full hardware speed for the most part, and feels nearly as responsive as it would if it were on the bare machine.

Figure 6.3 shows schematically how a virtual operating system runs on a host operating system; the guest operating system is an ordinary application as far as the host operating system is concerned.

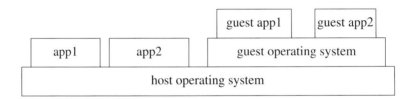

Figure 6.3: Virtual operating system organization.

Figure 6.4 is a screenshot of my Mac running VirtualBox, which in turn is running two guest operating systems: Linux on the left and Windows 10 on the right.

Cloud computing, which we will return to in Chapter 11, relies on virtual machines. A cloud service provider has a large number of physical computers, with plenty of storage and network bandwidth, which it uses to provide computing power to its customers. Each customer uses some number of virtual machines that are supported on fewer physical machines; multi-core processors are a natural fit for this kind of operation.

Amazon Web Services (AWS) is the largest provider of cloud computing, followed by Microsoft Azure and Google Cloud Platform; AWS has been particularly successful, accounting for well over half of Amazon's operating profit. These all offer a service whose capacity for any given customer can grow or shrink as load changes; there are sufficient resources to let individual users scale up or down instantaneously. Many companies, including large ones like Netflix, find cloud computing more cost-effective than running their own servers, thanks to economies of scale, adaptability to changing load, and less need for in-house staff.

Virtual operating systems raise some interesting ownership questions. If a company runs a large number of virtual Windows instances on one physical computer, how many Windows licenses does it need to buy from Microsoft? Ignoring legal issues, the answer is one, but Microsoft's licenses for Windows limit the total number

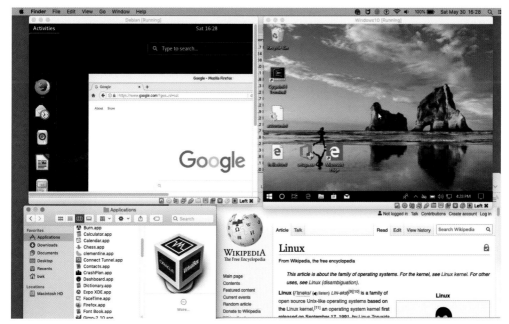

Figure 6.4: macOS running Windows and Linux virtual machines.

of virtual instances that you can legitimately run without paying for more copies.

Another use of the word "virtual" should be mentioned here. A program that simulates a computer, whether a real one or a pretend one (like the Toy), is also often called a *virtual machine*. That is, the computer exists only as software, a program that mimics its behavior as if it were hardware.

Such virtual machines are common. Browsers have one virtual machine to interpret JavaScript programs, and may have another separate virtual machine for Java programs. There is also a Java virtual machine in Android cell phones. Virtual machines are used because it's easier and more flexible to write and distribute a program than to build and ship physical equipment.

6.2 How an Operating System Works

The processor is constructed so that when the computer is powered on, the processor starts by executing a few instructions stored in a permanent memory. Those instructions in turn read instructions from a small flash memory that contains enough code to read more instructions from a known place on a disk, a USB memory, or a network connection, which reads still more instructions, until finally enough has been loaded to do useful work. This getting-started process was originally called "bootstrapping," after the old expression of pulling oneself up by one's bootstraps, and is now just *booting*. The details vary, but the basic idea is the same—a few instructions are sufficient to find more, which in turn lead to still more.

Part of this process may involve querying the hardware to determine what devices are attached to the computer, for example whether there is a printer or a scanner.

Memory and other components are checked to verify that they are working correctly. The boot process may involve loading software components (drivers) for connected devices so the operating system can use them. All of this takes time, while we're waiting impatiently for the computer to start doing something useful. It's frustrating that although computers are far faster than they used to be, they may still take a minute or two to boot.

Once the operating system is running, it settles down to a fairly simple cycle, giving control in turn each application that is ready to run or that needs attention. If I am typing text in a word processor, checking my mail, surfing randomly, and playing music in the background, the operating system gives the processor's attention to each of these processes one after another, switching the focus among them as necessary. Each program gets a short slice of time, which ends when the program requests a system service or its allocated time runs out.

The system responds to events like the end of the music, the arrival of mail or a web page, or a keypress; for each, it does whatever is necessary, often relaying the fact that something has happened to the application that must take care of it. If I decide to rearrange the windows on my screen, the operating system tells the display where to put the windows, and it tells each application what parts of that application's window are now visible so the application can redraw them. If I quit an application with File | Exit or by clicking the little × in the upper corner of the window, the system notifies the application that it's about to die, so it has a chance to put its affairs in order, for instance by asking the user "Do you want to save the file?" The operating system then reclaims any resources that the program was using, and tells apps whose windows are now exposed that they have to redraw.

6.2.1 System calls

An operating system provides an interface between the hardware and other software. It makes the hardware appear to provide higher-level services than it really does, so programming is easier. In the jargon of the field, the operating system provides a *platform* upon which applications can be built. It's another example of abstraction, providing an interface or surface that hides irregularities and irrelevant details of implementation.

The operating system defines a set of operations or services that it offers to application programs, like storing data in a file or retrieving data from a file, making network connections, fetching whatever has been typed on the keyboard, reporting mouse movements and button clicks, and drawing on the display.

The operating system makes these services available in a standardized or agreed-upon fashion, and an application program requests them by executing a particular instruction that transfers control to a specific place in the operating system. The system does whatever the request implies, and returns control and the results to the application. These entry points into the system are called *system calls*, and their detailed specification defines what the operating system is. A modern operating system typically has a few hundred system calls.

6.2.2 Device drivers

A *device driver* is code that acts as a bridge between the operating system and a specific kind of hardware device like a printer or a mouse. Driver code has detailed knowledge of how to make a particular device do whatever it does—how to access motion and button information from a specific mouse or trackpad, how to make a disk read and write information on an integrated circuit or a spinning magnetic surface, how to make a printer put marks on paper, or how to make a specific wireless chip send and receive radio signals.

The driver insulates the rest of the system from the idiosyncrasies of particular devices—all devices of one kind, like keyboards, have basic properties and operations that the operating system cares about—and the driver interface lets the operating system access the device in a uniform way so it's easy to switch devices.

Consider a printer. The operating system wants to make standard requests: print this text at this position on the page, draw this image, move to the next page, describe your capabilities, report your status, and so on, in a uniform way that would apply to any printer. Printers differ in their capabilities, however, for example whether they support color, two-sided printing, multiple paper sizes, and the like, and also in the mechanics of how marks are transferred to paper. The driver for a given printer is responsible for converting operating system requests into whatever is needed to make the particular device do those tasks, for example converting color to grayscale for a black and white printer. In effect, the operating system makes generic requests to an abstract or idealized device and the driver implements them for its own hardware. You can see this mechanism if you have multiple printers for a given computer: the dialog boxes for printing offer different options for the different printers.

A general-purpose operating system will have many drivers; for example, Windows ships with drivers already installed for an enormous variety of devices that might be potentially used by consumers, and every device manufacturer has a web site from which new and updated drivers can be downloaded.

Part of the boot process is to load drivers for currently available devices into the running system; the more devices there are, the more time this will take. It's also normal for new devices to appear out of nowhere. When an external disk is plugged into a USB socket, the operating system recognizes a new device, determines that it is a disk, and loads a USB disk driver for subsequent communication. Normally there's no need to find a new driver; the mechanism is so standardized that the operating system already has what it needs and the specifics of driving the device are buried in a processor within the device itself.

Figure 6.5 illustrates the relationships among the operating system, system calls, drivers, and applications. The picture would be similar for a cell phone system like Android or iOS.

6.3 Other Operating Systems

The existence of ever cheaper and smaller electronics makes it feasible to include more of that hardware in a device; as a result, many devices have significant processing power and memory. Calling a digital camera "a computer with a lens" is not far

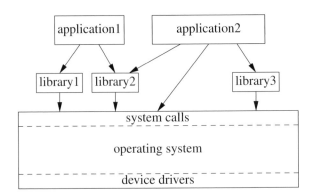

Figure 6.5: Operating system, system call, and device driver interfaces.

off the mark. As processing power and memory have increased, cameras have ever more capabilities; my inexpensive point-and-shoot camera records high-definition videos and uses Wi-Fi to upload pictures and videos to a computer or phone. Phones themselves are another great example, and of course cameras and phones are converging; any random phone today has far more megapixels than my first digital camera did, though the lens quality is another story.

The overall result is that devices are taking on the trappings of mainstream general-purpose computers like those we discussed in Chapter 1. They have a powerful processor, a lot of memory, and some peripheral devices, like the lens and display on a camera. They may have sophisticated user interfaces. They often have a network connection so they can talk to other systems—cell phones use the telephone network and Wi-Fi, while game controllers use infrared and Bluetooth—and many use USB for occasional ad hoc connections. The "Internet of Things" is based on this as well: thermostats, lights, security systems, and the like are controlled by embedded computers and are connected to the Internet.

As this trend continues, it increasingly makes more sense to use a commodity operating system than to write one's own. Unless the environment is unusual, it's easier and cheaper to use a stripped-down version of Linux, which is robust, adaptable, portable and free, rather than to develop one's own specialized system or license a costly commercial product. A drawback is that one may have to release some of the resulting code under a license like the GPL. This could raise issues of how to protect intellectual property in the device, but it has not proven insurmountable for the likes of Kindle and TiVo, along with many others.

6.4 File Systems

The file system is the part of the operating system that makes physical storage media like disks, CDs and DVDs, and other removable memory devices look like hierarchies of files and folders. The file system is a great example of the distinction between logical organization and physical implementation: file systems organize and store information on many different kinds of devices but the operating system presents the same interface for all of them. The ways that file systems store information can have practical and even legal implications, so another reason for studying file systems is to understand why "removing a file" does not mean that its contents are gone forever.

Most readers will have used Windows File Explorer or macOS Finder, which show the hierarchy starting from the top (the C: drive on Windows, for example). A *folder* contains the names of other folders and files; examining a folder will reveal more folders and files. (Unix systems traditionally use the word *directory* instead of folder.) The folders provide the organizational structure, while the files hold the actual contents of documents, pictures, music, spreadsheets, web pages, and so on. All the information that your computer holds is stored in the file system and is accessible through it if you poke around. This includes not only your data, but the executable forms of programs like Word and Chrome, libraries, configuration information, device drivers, and the files that make up the operating system itself. There's an astonishing amount; I was amazed to discover that my modest MacBook has over 900,000 files; a friend reports over 800,000 on one of his Windows computers. Figure 6.6 shows parts of five levels of hierarchy on my computer, leading down to some pictures in my home directory.

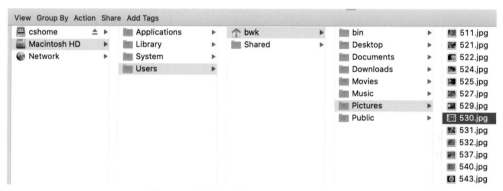

Figure 6.6: File system hierarchy.

In spite of their names, Finder and Explorer are most useful when you already know where your files are: you can always navigate from the root or top of the file system hierarchy. If you don't know where a file is located, however, you might have to use a search tool, like Spotlight on macOS.

The file system manages all this information, making it accessible for reading and writing by applications and the rest of the operating system. It coordinates accesses so they are performed efficiently and don't interfere with each other, it keeps track of where data is physically located, and it ensures that the pieces are kept separate so that parts of your email don't mysteriously wind up in your spreadsheets or tax returns. On systems that support multiple users, it enforces information privacy and security, making it impossible for one user to access another user's files without permission, and it may impose quotas on the amount of space that each user can use.

File system services are available through system calls at the lowest level, usually supplemented by software libraries to make common operations easy to program.

6.4.1 Secondary storage file systems

The file system is a wonderful example of how a wide variety of physical systems can be made to present a uniform logical appearance, a hierarchy of folders and files. How does it work?

A 500 GB drive holds 500 billion bytes, but software on the drive itself is likely to present this as something like 500 million chunks or *blocks* of 1,000 bytes each. (In real computers, these sizes would be powers of 2; I'm using decimal numbers to make it easier to see relationships.) A file of say 2,500 bytes, like a small mail message, would be stored in three of these blocks; it's too big for two blocks but three is enough.

File systems don't store bytes of one file in the same block as bytes of a different file, so there's some waste if the last block is not completely full: 500 bytes are unused in the last block in our example. That's a modest price for a considerable simplification in bookkeeping effort, especially since secondary storage is so cheap.

A folder entry for this file would contain its name, its size of 2,500 bytes, the date and time it was created or changed, and other miscellaneous facts about it (permissions, type, etc., depending on the operating system). All of that information is visible through a program like Explorer or Finder.

The folder entry also contains information about where the file is stored on the drive—which of the 500 million blocks contain its bytes. There are many different ways to manage that location information. The folder entry could contain a list of block numbers; it could refer to a block that itself contains a list of block numbers; or it could contain the number of the first block, which in turn gives the second block, and so on.

Figure 6.7 sketches an organization with blocks that refer to lists of blocks, as it might look on a conventional hard drive. Blocks need not be physically adjacent on a hard drive, and in fact they typically won't be, at least for large files. A megabyte file will occupy a thousand blocks, and those are sure to be scattered to some degree. The folders and the blocklists are themselves stored in blocks on the same drive, though that's not shown in the diagram.

The physical implementation would be very different for a solid state drive, but the basic idea is the same. As noted earlier, most computers today use SSDs because although more expensive per byte, they're smaller and offer greater reliability, lower weight, and lower power consumption. Viewed from a program like Finder or Explorer, there's no difference whatsoever. But an SSD device will have a different driver, and the device itself has sophisticated code to remember where information is located on the device. This is because SSD devices are limited by how many times each part can be used. Software in the drive keeps track of how much each physical block has been used and moves data around to ensure that each block is used roughly the same amount, a process called *wear leveling*.

A folder is a file that contains information about where folders and files are located. Because information about file contents and organization must be perfectly accurate and consistent, the file system reserves to itself the right to manage and maintain the contents of folders. Users and application programs can only change

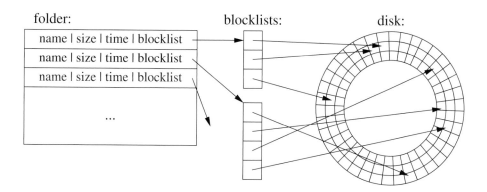

Figure 6.7: File system organization on a hard disk.

folder contents implicitly, by making requests of the file system.

From one perspective, folders *are* files; there's no difference in how they are stored except that the file system is totally responsible for folder contents, and application programs have no direct way to change them. But at the lowest level it's just blocks, all managed by the same mechanisms.

When a program wants to access an existing file, the file system has to search for the file starting at the root of the file system hierarchy, looking for each component of the file path name in the corresponding folder. That is, if the file is /Users/bwk/ book/book.txt on a Mac, the file system will search the root of the file system for Users, then search within that folder for bwk, then within that folder for book, then within that for book.txt. On Windows, the name might be C:\My Documents\book\book.txt, and the search would be analogous.

This is an efficient strategy, since each component of the path narrows the search to files and folders that lie within that folder; all others are eliminated. Thus multiple files can have the same name for some component; the only requirement is that the full path name be unique. In practice, programs and the operating system keep track of the folder that is currently in use, so searches need not start from the root each time, and the system will also cache frequently used folders to speed up operations.

When a program wants to create a new file, it makes a request of the file system, which puts a new entry in the appropriate folder, including name, date, and so on, and a size of zero (since no blocks are allocated to the brand new file yet). When the program later writes data into the file, say by appending the text of a mail message, the file system finds enough currently unused or "free" blocks to hold the requested information, copies the data into them, inserts them into the folder's list of blocks, and returns to the application.

This suggests that the file system maintains a list of all the blocks on the drive that are not currently in use, that is, not already part of some file. When a request for a new block arrives, it can be satisfied with a block taken from the list of free blocks. The free list is kept in file system blocks too, but is only accessible to the operating system, not to application programs.

6.4.2 Removing files

When a file is removed, the opposite happens: the blocks of the file are returned to the free list and the folder entry for the file can be cleared, so the file appears to have gone away. In reality, it's not quite like that, and therein lie a number of interesting implications.

When a file is removed in Windows or macOS it goes to the "Recycle Bin" or "Trash," which appears to be just another folder, albeit with slightly different properties. Indeed, that's exactly what the recycle bin is. When a file is to be removed, its folder entry and full name is copied from the current folder to a folder called Recycle or Trash, and the original folder entry is cleared. The blocks of the file and thus its contents do not change at all! Recovering a file from the recycle bin reverses the process, restoring the entry to its original folder.

"Emptying the trash" is more like what we originally described, where the Recycle or Trash folder entry is cleared and the blocks are really added to the free list. This is true whether the emptying is done explicitly or happens quietly behind your back because the file system knows it's running low on free space.

Suppose you empty the trash explicitly by clicking "Empty recycle bin" or "Empty trash." That clears the entries in the recycle folder itself and puts the blocks on the free list, but their contents haven't yet been deleted—all the bytes of each block of the original files are still sitting there untouched. They won't be overwritten with new content until that block is taken off the free list and given to a new file.

This delay means that information that you thought was removed still exists and is readily accessible if one knows how to find it. Any program that reads the drive by physical blocks, that is, without going through the file system hierarchy, can see what the old content was. In mid-2020, Microsoft announced Windows File Recovery, a free tool for doing exactly this kind of recovery for a wide variety of file systems and media.

This has potential benefits. If something goes wrong with your disk, it might still be possible to recover information even if the file system is confused. There's no guarantee that the data is really gone, however, which is bad if you truly want the information removed, perhaps because it's private or you are plotting some evil deed. A competent enemy or law enforcement agency would have no trouble recovering it. If you're planning something nefarious or if you're just paranoid, you have to use a program that will erase the information from freed blocks.

In practice you may have to do better than this, since a truly dedicated adversary with lots of resources might be able to extract traces of information even when new information has overwritten it. Military-grade file erasing overwrites blocks multiple times with random patterns of 1s and 0s. Even better is to demagnetize a hard disk by putting it near a strong magnet. Best of all is to physically destroy it; that's the only way to ensure that the contents are gone.

Even that might not be enough, however, if your data is being backed up automatically all the time (as mine is at work) or if your files are kept on a network file system or somewhere "in the cloud" rather than on your own drive. (And if you sell or give away an old computer or phone, you might want to ensure that any data on it is unrecoverable.)

A somewhat similar situation applies to the folder entry itself. When you remove a file, the file system will note that the folder entry no longer points to a valid file. It could do that by setting a bit in the folder that means "this entry is not in use." Then it would be possible to recover the original information about the file, including the contents of any blocks that had not been re-allocated, until the folder entry was itself re-used. This mechanism was at the heart of commercial file recovery programs for Microsoft's MS-DOS system in the 1980s, which marked free entries by setting the first character of the filename to a special value. This made it easy to recover the whole file if the recovery was attempted soon enough.

The fact that the contents of files can survive long after their creator thought them deleted has implications for legal procedures like discovery and document retention. It's remarkably common, for example, for old email messages to surface that are in some way embarrassing or incriminating. If records exist only on paper, there's a decent chance that careful shredding will destroy all copies, but digital records proliferate, are readily copied onto removable devices, and can be tucked away in many places. The results of searching the web for phrases like "emails reveal" or "leaked email" should convince you to be circumspect about what you say in mail and indeed in any information that you commit to a computer.

6.4.3 Other file systems

I've been discussing conventional file systems on secondary storage drives, since that's where much of the information is and that's what we most often see on our own computers. But the file system abstraction applies to other media as well.

CD-ROMs and DVDs provide access to their information as if it were a file system, again with a hierarchy of folders and files. Flash memory file systems on USB and SD ("Secure Digital," Figure 6.8) drives are ubiquitous. When plugged into a Windows computer, a flash drive appears as another disk drive. It can be explored with File Explorer and files can be read and written exactly as if it were built-in. The only difference is that its capacity might be smaller and access may be somewhat slower.

Figure 6.8: SD card flash memory.

If that same device is plugged into a Mac, it appears there as a folder as well, to be explored with Finder, and files can be transferred back and forth. It can also be plugged into Unix or Linux computers, and again appears in the file system on those systems. Software makes the physical device look like a file system, with the same abstractions of folders and files, on a variety of operating systems. Internally, the organization is likely to be a Microsoft FAT file system, the widely used *de facto* standard, but we don't know for sure and we don't need to. The abstraction is perfect. ("FAT" stands for File Allocation Table; it's not a commentary on implementation quality.) Standardization of hardware interfaces and software structure makes this possible.

My first digital camera stored pictures in an internal file system and I had to connect the camera to a computer and run proprietary software to retrieve them. Every camera since has had a removable SD memory card, like the one shown in Figure 6.8, and I can upload pictures by moving the card from camera to computer. This is much faster than it was before, and as an unexpected fringe benefit, it has freed me from the camera manufacturer's appallingly awkward and flaky software. A familiar and uniform interface with standard media replaces clumsy and unique software and hardware. I imagine that the manufacturer is also happy that it's no longer necessary to provide specialized file transfer software.

It's worth mentioning one other version of the same idea: the network file system, which is common in schools and businesses. Software makes it possible to access the file system on some other computer as if it were on one's own machine, again using File Explorer, Finder or other programs to access the information. The file system at the far end may be of the same kind (both Windows computers, for example) or it might be something different, say macOS or Linux. As with flash memory devices, the software hides the differences and presents a uniform interface so it looks like a regular file system on the local machine.

Network file systems are often used as backup as well as primary file storage. Multiple older copies of files can be copied onto archival media for storage at a different site; this protects against a disaster like a ransomware attack or a fire that destroys the only copy of vital records. Some disk systems also rely on a technique called RAID ("redundant array of independent disks") that writes data onto multiple disks with an error-correction algorithm that makes it possible to recover information even if one of the disks breaks. Naturally such systems also make it hard to ensure that all traces of information have been erased.

Cloud computing systems, which we will discuss further in Chapter 11, have some of the same properties but typically don't present their contents with a file system interface.

6.5 Applications

"Application" is a blanket term for all kinds of programs or software systems that do some task, using the operating system as a platform. An application might be tiny or large; it might focus on one specific task or handle a broad range; it might be sold or given away; its code could be highly proprietary, freely available open source, or unrestricted.

Applications come in a wide range of sizes, from small self-contained programs that do only one thing to large programs that do complex sets of operations, like Word or Photoshop.

As an example of a simple application, consider the Unix program called `date`, which prints the current date and time:

```
$ date
Fri Nov 27 16:50:00 EST 2020
```

The `date` program behaves the same way on Unix-like systems, including macOS and is similar on Windows. The implementation of `date` is tiny, because it builds on a system call (`time`) that provides the current date and time in an internal format, and on libraries for formatting dates (`ctime`) and printing text (`printf`). Here's a complete implementation in C so you can see how short it is:

```
#include <stdio.h>
#include <time.h>
int main() {
    time_t t = time(0);
    printf("%s", ctime(&t));
    return 0;
}
```

Unix systems have a program called `ls` that lists the files and folders in a directory, a bare-bones text-only analog of programs like Windows File Explorer and macOS Finder. Other programs copy files, move them, rename them, and so on, operations that have graphical equivalents in Finder and Explorer. Again, these programs use system calls to access the basic information about what is in folders, and rely on libraries to read, write, format and display information.

An application like Word is much, much bigger than a program to explore the file system. It clearly has to include some of the same kind of file system code so that users can open files, read their contents and save documents in the file system. It includes sophisticated algorithms, for example to update the display continuously as the text changes. It supports an elaborate user interface that displays information and provides ways to adjust sizes, fonts, colors, layout, and so on; this is likely to be a major part of the program. Word and other large programs of substantial commercial value undergo continuous evolution as new features are added. I have no idea how big the source code for Word is, but I would not be surprised if it were ten million lines of C, C++ and other languages, especially if one includes variants for Windows, Macs, phones and browsers.

A browser is an example of a large, free, and sometimes open source application of even greater complexity in some dimensions. You have surely used at least one of Firefox, Safari, Edge or Chrome, and many people routinely use several. Chapter 10 will talk more about the web and how browsers fetch information; here I want to focus on the ideas in big, complicated programs.

Seen from the outside, a browser sends requests to web servers and retrieves information from them for display. Where's the complexity?

First, the browser has to deal with *asynchronous* events, that is, events that happen at unpredictable times and in no particular order. For instance, the browser has sent a request for a page because you clicked on a link, but it can't just wait for the reply; it has to remain responsive in case you scroll the current page, or abort the request if you hit the Back button or click on a different link, even while the requested page is coming in. It has to update the display if you reshape the window, perhaps continuously as you reshape back and forth while data is arriving. If the page includes sound or movies, the browser has to manage those as well. Programming asynchronous systems is always hard, and browsers must deal with lots of asynchrony.

The browser has to support many kinds of content, from static text to interactive programs that want to change what the page contains. Some of this can be delegated to helper programs—this is the norm for standard formats like PDF and movies—but the browser has to provide mechanisms for starting such helpers, sending and receiving data and requests for them, and integrating them into the display.

The browser manages multiple tabs and/or multiple windows, each of which may be doing some of the foregoing operations. It maintains a history for each of these, along with other data like bookmarks, favorites, and so on. It accesses the local file system for uploads, downloads and caching images.

It provides a platform for extensions at several levels: plug-ins like QuickTime, a virtual machine for JavaScript, and add-ons like Adblock Plus and Ghostery. Underneath, it has to work on multiple versions of multiple operating systems, including mobile devices.

With all of this complicated code, a browser is vulnerable to attacks through bugs in its own implementation or in the programs it enables, and through the innocence, ignorance and ill-advised behavior of its users, most of whom (save the readers of this book) have almost no understanding of what's going on or what the risks might be. It's not easy.

If you look back over the description in this section, does it remind you of something? A browser is similar to an operating system. It manages resources, it controls and coordinates simultaneous activities, it stores and retrieves information from multiple sources, and it provides a platform on which application programs can run.

For many years, it has seemed like it should be possible to use the browser as the operating system and thus be independent of whatever operating system is controlling the underlying hardware. A decade or two ago, this was a fine idea but there were too many practical hurdles. Today it's a viable alternative. Numerous services are already accessed exclusively through a browser interface—mail, calendars, music, videos and social networks are obvious examples—and this will continue. Google offers an operating system called Chrome OS that relies primarily on web-based services. A Chromebook is a computer that runs Chrome OS; it has only a limited amount of local storage, using the web for most storage, and it only runs browser-based applications like Google Docs. We'll come back to this topic when we talk about cloud computing in Chapter 11.

6.6 Layers of Software

Software, like many other things in computing, is organized into layers, analogous to geological strata, that separate one concern from another. Layering is one of the important ideas that help programmers to manage complexity. Each layer implements something, and provides an abstraction that the layer above can use for access to services.

At the bottom, at least for our purposes, is the hardware, which is more or less immutable except that buses make it possible to add and remove devices even while the system is running.

The next level is the operating system proper, often called the *kernel* to suggest its central function. The operating system is a layer between the hardware and the applications. No matter what the hardware, the operating system can hide its specific properties and provide applications with an interface or facade that is independent of many of the details of the specific hardware. If the interface has been well designed, it's possible for the same operating system interface to be available on different kinds of processors and to be provided by different suppliers.

This is true of the Unix and Linux operating system interface—Unix and Linux run on all kinds of processors, providing the same operating system services on each. In effect, the operating system has become a commodity; the underlying hardware doesn't matter much except for price and performance, and the software on top doesn't depend on it. (One way this is evident is that I'll often use "Unix" and "Linux" interchangeably, since for most purposes the distinction is irrelevant.) With care, all that's necessary to move a program to a new processor is to compile it with a suitable compiler. Of course the more tightly a program is tied to particular hardware properties, the harder this job will be, but it's eminently feasible for many programs.

As a large-scale example, Apple converted its software from the IBM PowerPC processor to Intel processors in less than a year in 2005–2006. In mid-2020, Apple announced that it was going to do the same thing again, henceforth using ARM processors in all its phones, tablets and computers, rather than processors from Intel. This is another demonstration of how software can be largely independent of a specific processor architecture.

This has been less true of Windows, which for many years was fairly closely tied to the Intel architecture that began with the Intel 8086 processor in 1978 and its many evolutionary steps since. (The family of processors is often called "x86" since for many years Intel processors had numbers ending in 86, including the 80286, 80386, and 80486.) The association was so tight that Windows running on Intel was sometimes called "Wintel." Today, however, Windows also runs on ARM processors.

The next layer above the operating system is a set of libraries that provide generally useful services so that individual programmers don't have to re-create them; these are accessed through their APIs. Some libraries are at a low level, dealing with basic functionality (computing mathematical functions like square root or logarithm, for example, or date and time computations like those in the `date` command above); others are much more complicated (cryptography, graphics, compression). Components for graphical user interfaces—dialog boxes, menus, buttons, check boxes, scroll bars, tabbed panes, and the like—involve a lot of code; once they are in a library, everyone can use them, which helps to ensure a uniform look and feel. That's why most Windows applications, or at least their basic graphical components, look so similar; the same is even more true on a Mac. It's too much work for most software vendors to re-invent and re-implement, and pointlessly different visual appearances are confusing to users.

Sometimes the distinction between kernel, library and application is not as clear as I have made it sound, since there are many ways to create and connect software components. For instance, the kernel could provide fewer services and rely on libraries in a layer above to do most of the work. Or it could take on more of the task itself, relying less on libraries. The border between operating system and application is not sharply defined.

What is the dividing line? A useful guideline, though not perfect, is that anything necessary to ensure that one application does not interfere with another is part of the operating system. Memory management—deciding where to put programs in RAM as they run—is part of the operating system. Similarly, the file system—where to store information on secondary storage—is a central function. So is control of devices—two applications should not be able to run the printer at the same time, nor write to the display without coordination. At the core, control of the processors is an operating system function, since that is necessary to ensure all the other properties.

A browser is not part of the operating system, since it's possible to run any browser, or multiple ones simultaneously, without interfering with shared resources or control. This might sound like a technical fine point, but it has had major legal ramifications. The Department of Justice versus Microsoft antitrust lawsuit that began in 1998 and ended in 2011 was in part about whether Microsoft's Internet Explorer browser ("IE") was part of the operating system or merely an application. If it was part of the system, as Microsoft argued, then it could not reasonably be removed and Microsoft was within its rights to require the use of IE. If it was just an application, however, then Microsoft could be deemed to be illegally forcing others to use IE when they did not need to. The case was of course more complicated than this, but the dispute about where to draw this line was an important part. For the record, the court decided that a browser is an application, not part of the operating system; in the words of Judge Thomas Jackson, "Web browsers and operating systems are separate products."

6.7 Summary

Applications get things done, with the operating system acting as coordinator and traffic cop to ensure that applications share resources—processor time, memory, secondary storage, network connections, other devices—efficiently and equitably, and do not interfere with each other. Essentially all computers today have an operating system, and the trend is towards using a general-purpose system like Linux rather than something specialized, because unless there are unusual circumstances, it's easier and cheaper to use existing code than to write new.

Much of the discussion in this chapter has been phrased in terms of applications for individual consumers, but many large software systems are invisible to most of their users. These include the programs that operate infrastructure like telephone networks, power grids, transportation services, and financial and banking systems. Planes and air traffic control, cars, medical devices, weapons, and so on are all run by large software systems. Indeed, it's hard to think of any significant technology that we use today that doesn't have a major software component.

Software systems are big, complicated, and often buggy, and all of these are made worse by constant change. It's difficult to get accurate estimates of how much code there is in any big system, but the major systems that we rely on tend to involve millions of lines at a minimum. Thus it's inevitable that there will be significant bugs that can be exploited. As our systems get more complicated, this situation is likely to get worse, not better.

7

Learning to Program

"Don't just play on your phone, program it!"

    President Barack Obama, December 2013.

In my course, I teach a small amount of programming, because I think that it's important for a well-informed person to know something about programming, perhaps only that it can be surprisingly difficult to get even simple programs to work properly. There is nothing like doing battle with a computer to teach this lesson, but also to give people a taste of the wonderful feeling of accomplishment when a program does work for the first time. It may also be valuable to have enough programming experience that you are cautious when someone says that programming is easy or that there are no errors in a program. If you had trouble making 10 lines of code work after a day of struggle, you might be legitimately skeptical of someone who claims that a million-line program will be delivered on time and bug-free. On the other hand, there are times when it's helpful to know that not all programming tasks are difficult, for instance when hiring a consultant.

There are zillions of languages. Which one should you learn first? If you want to program your phone, as President Obama exhorted us, you need Java for Android or Swift for iPhones; both can be learned by beginners but are difficult for casual use, and phone programming has a lot of details as well. Scratch, a visual programming system from MIT, is especially good for children, but it does not scale up to larger or more complicated programs.

In this chapter, I'm going to talk briefly about two programming languages, JavaScript and Python. Both are widely used by both amateur and professional programmers. They are easy to learn at a beginner level, scale up to larger programs, and are broadly applicable.

JavaScript is included in every browser so there's no software to download. If you do write a program, you can use it on your own web pages to show your friends and family. The language itself is simple and one can do neat things with comparatively little experience; at the same time, it is remarkably flexible. Almost every web page includes some JavaScript, and that code can be examined from within a browser by viewing the page source, though you will have to go through a couple of menus to find the right item, and browsers make it harder to find than they should. Many web page effects are enabled by JavaScript, including Google Docs and equivalent programs from other sources. JavaScript is also the language for APIs provided by web services like Twitter, Facebook, Amazon, and so on.

JavaScript has disadvantages too. Some parts of the language are awkward and there are some surprising behaviors. The browser interface is not as standardized as one would like, so programs don't always behave the same way on different browsers. At the level we're talking about, this is not an issue, and even for professional programmers it's getting better all the time.

JavaScript programs generally run as part of a web page, though non-browser use is growing. When JavaScript is used with a browser as host, one has to learn a small amount of HTML (Hypertext Markup Language), the language for describing the layout of web pages. (We will see a bit of that in Chapter 10.) In spite of these minor drawbacks, it's well worth the effort of learning a little JavaScript.

Our other language is Python. Python is excellent for day-to-day programming over an immense range of potential applications. In the past few years, Python has become a standard language for introductory programming classes, and for classes focused on data science and machine learning. Although you would normally run Python on your own computer, there are now web sites that make it possible to run Python programs as a web service so there's no need to download anything nor learn how to use a command-line interface. If I were teaching a programming course for people who were learning their first language, I would use Python.

If you follow the material here and do some experimentation, you can learn how to program, at least at a basic level, and that's a skill worth having. The knowledge you acquire will carry over into other languages, making them easier to learn. If you want to dig deeper or get a different take on it, search for JavaScript or Python tutorials on the web and you'll get a long list of helpful sites, including Codecademy, the Khan Academy, and W3Schools, that teach programming to absolute beginners.

All that said, however, it's fine to skim this chapter and ignore syntactic details; nothing else depends on it.

7.1 Programming Language Concepts

Programming languages share certain basic ideas, since they are all notations for spelling out a computation as a sequence of steps. Every programming language thus will provide ways to read input data, do arithmetic, store and retrieve intermediate values as computation proceeds, decide how to proceed on the basis of previous computations, display results along the way, and save results when the computation is finished.

Languages have *syntax*, that is, rules that define what is grammatically legal and what is not. Programming languages are picky about grammar: you have to say it right or there will be a complaint. Languages also have *semantics*, that is, a well-defined meaning for anything you can say in the language.

In theory there should be no ambiguity about whether a particular program is syntactically correct and, if so, what its meaning is. This ideal is not always achieved, unfortunately. Languages are usually defined in words and, like any other document in a natural language, the definitions can have ambiguities and allow for different interpretations. On top of this, implementers can make mistakes, and languages evolve over time. Accordingly, JavaScript implementations differ somewhat from browser to browser, and even from version to version of the same browser. Similarly, there are two versions of Python, largely compatible but with just enough differences to be irritating. Fortunately, version 2 is on the way out, being replaced by version 3, and this will cease to be a problem.

Most programming languages have three aspects. First is the language itself: statements that tell the computer to do arithmetic, test conditions, and repeat computations. Second, there are libraries of code that others have written that you can use in your own programs; these are prefabricated pieces that you don't have to write for yourself. Typical examples include math functions, calendar computations, and functions for searching and manipulating text. Third, there is access to the environment in which the program runs. A JavaScript program running in a browser can get input from a user, react to events like button pushes or typing into a form, and cause the browser to display different content or go to a different page. A Python program can access the file system on the computer where it is running, something that a JavaScript program running in a browser is prohibited from doing.

7.2 A First JavaScript Program

I'm going to start with JavaScript, then follow up with Python. The ideas from the JavaScript part will make it easier to read the Python sections, but you could read in the opposite order as well. Generally, after you learn one language, others come easily because you understand the concepts and only have to pick up a new syntax.

The first JavaScript program is as small as they get: it just pops up a dialog box that says "Hello, world" when the web page is loaded. Here's the complete page in HTML, which we will meet when we talk about the World Wide Web in Chapter 10. For now focus on the single line of highlighted JavaScript code, which appears between <script> and </script>.

```
<html>
  <body>
    <script>
      alert("Hello, world");
    </script>
  </body>
</html>
```

If you put these seven lines into a file called hello.html and load that file into your browser, you'll see a result like one of those in Figure 7.1.

The images are from Firefox, Chrome, Edge and Safari on macOS. You can see that different browsers may behave differently. Notice that Safari displays "Close" but not as a button; Edge is nearly identical to Chrome because Edge is built on the Chrome implementation.

The alert function is part of the JavaScript library for interacting with the browser. It pops up a dialog box that displays whatever text appears between the

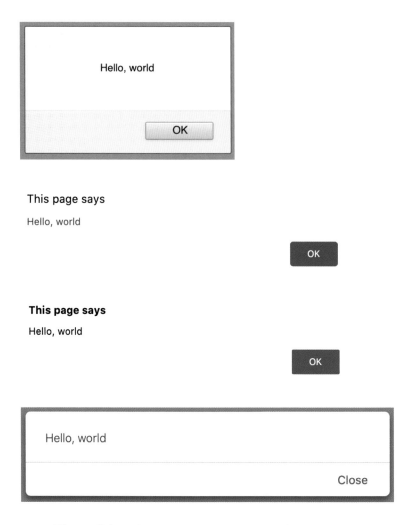

Figure 7.1:   Firefox, Chrome, Edge, Safari, all on macOS.

quotes and waits for the user to push OK or Close. By the way, when you create your own JavaScript programs, you must use the standard double quote character that looks like this ", not the so-called "smart quotes" that you see in ordinary text. This is a simple example of a syntax rule. Don't use a word processor like Word to create HTML files; use a text editor like Notepad or TextEdit and make sure that it saves files as plain text (that is, plain ASCII without formatting information) even when the filename extension is .html.

Once you have this example working, you can extend it to do more interesting computations. I won't show the HTML parts from now on, just the JavaScript, which goes between <script> and </script>.

7.3 A Second JavaScript Program

The second JavaScript program asks the user for a name, then displays a personalized greeting:

```
var username;
username = prompt("What's your name?");
alert("Hello, " + username);
```

This program has several new constructs and corresponding ideas. First, the word `var` introduces or *declares* a *variable*, which is a place in primary memory where the program can store a value as the program runs. It's called a variable because its value can be changed as a result of what the program is doing. Declaring a variable is the high-level language analog of giving a name to a memory location as we did in the Toy assembly language. Metaphorically, declarations specify the *dramatis personae*, the list of characters in a play. I named the variable `username`, which describes its role in this program.

Second, the program uses a JavaScript library function called `prompt`, which is similar to `alert` but pops up a dialog box that asks the user for input. Whatever text the user types is made available to the program as the value computed by the `prompt` function. That value is assigned to the variable `username` by the line

```
username = prompt("What's your name?");
```

The equals sign "=" means "perform the operation on the right side and store the result in the variable named on the left side," like storing the accumulator value in memory in the Toy; interpreting the equals sign this way is an example of semantics. This operation is called *assignment* and = does not mean equality; it means copying a value. Most programming languages use the equals sign for assignment, in spite of potential confusion with mathematical equality.

Finally, the plus sign + is used in the `alert` statement

```
alert("Hello, " + username);
```

to join the word `Hello` (and a comma and a space) and the name that was provided by the user. This is potentially confusing too because + in this context does not mean addition of numbers but concatenation of two strings of text characters.

When you run this program, `prompt` displays a dialog box where you can type something, shown in Figure 7.2 (from Firefox).

If you type "Joe" into this dialog box and then push OK, the result is the message box seen in Figure 7.3.

It would be an easy extension to allow for a first name and a last name as separate inputs, and there are plenty of variants that you might try for practice. Notice that if you respond with "My name is Joe", the result will be "Hello, My name is Joe". If you want smarter behavior from a computer, you have to program it yourself.

Figure 7.2: Dialog box waiting for input.

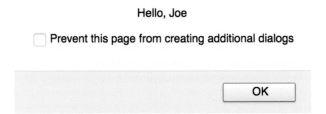

Figure 7.3: Result of responding OK to dialog box.

7.4 Loops and Conditionals

Figure 5.6 is a JavaScript version of the program that adds up a sequence of numbers. Figure 7.4 shows it again so you don't have to look back.

```
var num, sum;
sum = 0;
num = prompt("Enter new value, or 0 to end");
while (num != '0') {
    sum = sum + parseInt(num);
    num = prompt("Enter new value, or 0 to end");
}
alert(sum);
```

Figure 7.4: JavaScript program to add up numbers.

As a reminder, the program reads numbers until a zero is entered, then prints the sum. We've already seen several of the language features in this program, like declarations, assignments, and the `prompt` function. The first line is a variable declaration that names the two variables `num` and `sum` that the program will use. The second line is an assignment statement that sets `sum` to zero, and the third line sets `num` to the value that the user types into the dialog box.

The important new feature is the `while` loop, which includes lines 4 through 7. Computers are wonderful devices for repeating sequences of instructions over and over again; the question is how to express that repetition in a programming language. The Toy language introduced the `GOTO` instruction, which branches to another place

in a program rather than the next instruction in sequence, and the IFZERO instruction, which branches only if the accumulator value is zero.

These ideas appear in most high-level languages in a statement called a *while
loop*, which provides a more orderly and disciplined way to repeat a sequence of
operations. The while statement tests a condition (written between parentheses)
and if the condition is true, executes all the statements between the { ... } braces in
order. It then goes back and tests the condition again. This cycle continues while the
condition is true. When the condition becomes false, execution continues with the
statement that follows the closing brace of the loop.

This is almost an exact match for what we wrote with IFZERO and GOTO in Toy
programs in Chapter 3, except that with a while loop we don't need to invent labels
and the test condition can be any expression that evaluates to true or false. Here the
test is whether the variable num is not the character 0. The operator != means "not
equal to"; it's inherited from C, as is the while statement itself.

I've been casual about the type of data that these sample programs process, but
internally computers make a strong distinction between numbers like 123 and arbitrary text like Hello. Some languages require programmers to express that distinction carefully; other languages try to guess what the programmer might have meant.
JavaScript is closer to the latter position, so sometimes it's necessary to be explicit
about the type of data you're dealing with and how values are to be interpreted.

The function prompt returns characters (text) and the test determines whether
the returned characters are a literal 0, which is expressed by placing it in quotes.
Without the quotes, it would be a *numeric* zero.

The function parseInt converts text into an internal form that can be used for
integer arithmetic. In other words, its input data is to be treated as an integer (like
123) instead of as three characters that happen to be decimal digits. If we don't use
parseInt, the data returned by prompt will be interpreted as text and the + operator will append it to the end of the previous text. The result would be the concatenation of all the digits the user entered—interesting, perhaps, but not what we intended.

The next example, in Figure 7.5, does a slightly different job, finding the numerically largest of all the numbers that were entered. It's an excuse to introduce another
control-flow statement, if-else, which appears in some form in all high-level languages as a way to make decisions. In effect, it's a general-purpose version of
IFZERO. JavaScript's version of if-else is the same as in C.

The if-else statement comes in two forms. The one shown here has no else
part: if the parenthesized condition is true, the statements in { ... } braces that follow
are executed. Regardless, execution continues with the statement that follows the
closing brace. The more general form has an else part for a sequence of statements
to be executed if the condition is false. Either way, execution continues with the next
statement after the whole if-else.

You might have noticed that the example programs use indentation to highlight
structure: the statements controlled by while and if are indented. This is good
practice because it makes it possible to see at a glance what the scope is for statements like while and if that control other statements.

It's easy to test this program if you run it from a web page, but professional programmers would check it out even before that, simulating its behavior by mentally

```
var max, num;
max = 0;
num = prompt("Enter new value, or 0 to end");
while (num != '0') {
   if (parseInt(num) > parseInt(max)) {
      max = num;
   }
   num = prompt("Enter new value, or 0 to end");
}
alert("Maximum is " + max);
```

Figure 7.5: Finding the largest of a sequence of numbers.

stepping through the statements of the program one at a time, doing what a real computer would do. For example, try the input sequences 1, 2, 0 and 2, 1, 0; you might even start with the sequences 0 and then 1, 0 to be sure the simplest cases work properly. If you do that (it's good practice for being sure you understand how it works), you will conclude that the program works for any sequence of input values.

Or does it? It works fine if the inputs include at least one positive number, but what happens if all of them are negative? If you try that, you'll discover that the program always says that the maximum is zero.

Think about why for a moment. The program keeps track of the largest value seen so far in a variable called max (just like finding the tallest person in the room). The variable has to have some initial value before we can compare subsequent numbers to it, so the program sets it to zero at the beginning, before any numbers have been provided by the user. That's fine if at least one input value is greater than zero, as would be the case for heights. But if all inputs are negative, the program doesn't print the largest of the negative values; instead it prints the original value of max, which is never updated.

This bug is easy to eliminate. I'll show one solution at the end of the JavaScript discussion, but it's a good exercise to discover a fix by yourself.

The other thing that this example shows is the importance of testing. Testing requires more than throwing random inputs at a program. Good testers think hard about what could go wrong, including weird or invalid input, and "edge" or "boundary" cases like no data at all or division by zero. A good tester would think of the possibility of all negative inputs. The problem is that as programs become larger, it's harder and harder to think of all the test cases, especially when they involve humans who are likely to input random values at random times in random order. There's no perfect solution, but careful program design and implementation helps, as does including consistency and sanity checks in the program from the beginning, so that if something does go wrong, it's likely to be caught early by the program itself.

7.5 JavaScript Libraries and Interfaces

JavaScript has an important role as the extension mechanism for sophisticated web applications. Google Maps is a nice example. It provides a library and an API so that map operations can be controlled by JavaScript programs, not just by mouse clicks. Thus anyone can write JavaScript programs that display information on a map provided by Google. The API is easy to use; for example the code in Figure 7.6 (with a few extra lines of HTML and an authorization key from Google) displays the map image of Figure 7.7, where perhaps some reader of this book will live some day.

```
function initMap() {
  var latlong = new google.maps.LatLng(38.89768, -77.0365);
  var opts = {
    zoom: 18,
    center: latlong,
    mapTypeId: google.maps.MapTypeId.HYBRID
  };
  var map = new google.maps.Map(
               document.getElementById("map"), opts);
  var marker = new google.maps.Marker({
    position: latlong,
    map: map,
  });
}
```

Figure 7.6: JavaScript code to use Google Maps.

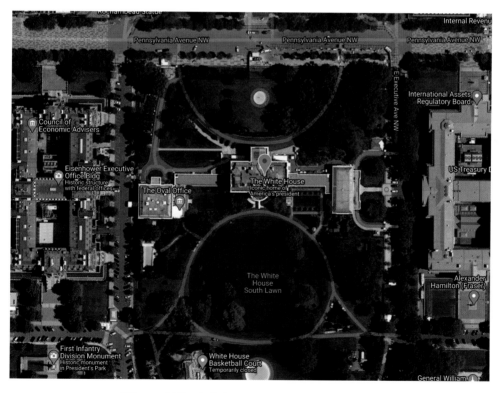

Figure 7.7: Might you live here some day?

As we'll see in Chapter 11, the trend on the web is towards more and more use of JavaScript, including programmable interfaces like Maps. One downside of this is that it's difficult to protect intellectual property when you are forced to reveal source code, which you necessarily must do if you're using JavaScript. Anyone can use the browser to look at the source of a page. Some JavaScript is obfuscated, either intentionally or as a by-product of trying to make it compact so that it can be downloaded more quickly, and the result can be totally impenetrable unless one is determined.

7.6 How JavaScript Works

Recall the discussion of compilers, assemblers and machine instructions in Chapter 5. A JavaScript program is converted into executable form in an analogous way, although there are significantly different details. When the browser encounters JavaScript in a web page (when it sees a `<script>` tag, for example), it passes the text of the program to a JavaScript compiler. The compiler checks the program for errors, and compiles it into assembly language instructions for a made-up machine analogous to the Toy, though with a richer instruction repertoire: a virtual machine as described in the previous chapter. It then runs a simulator like the Toy to perform whatever actions the JavaScript program is supposed to do. The simulator and the browser interact closely; when a user pushes a button, for example, the browser notifies the simulator that the button has been pushed. When the simulator wants to do something like pop up a dialog box, it asks the browser to do the work by calling `alert` or `prompt`.

This is all we'll say about JavaScript here, but if you're interested in more, there are good books, and online tutorials that let you edit JavaScript code in place and show you the results immediately. Programming can be frustrating, but it can also be great fun, and you can even make a decent living with it. Anyone can be a programmer, but it helps if you have an eye for detail and are able to zoom in to look at fine points and zoom out again to the big picture. It also helps to be a bit compulsive about getting details right, since programs won't work well or perhaps not at all if you're not careful. And as with most activities, there is a large gap between an amateur and a real professional.

Here's one possible answer to the programming question a few pages back:

```
num = prompt("Enter new value, or 0 to end");
max = num;
while (num != '0') ...
```

Set `max` to the first number that the user supplies; that is the largest value so far, no matter whether it's positive or negative. Nothing else needs to change, and the program now handles all inputs, though it will quit early if one of those values is zero. It even does something sensible if the user provides no values at all, though to handle that well in general requires learning more about the `prompt` function.

7.7 A First Python Program

I'm now going to reprise some of the material from the first part of the chapter in Python, focusing on differences with JavaScript. One major change from a few years ago is that it is now easy to run Python programs from a browser, which means that, as with JavaScript, there's no need to download anything to your own computer. There are still restrictions on what you can access and what resources are available, since you are running your program on someone else's computer, but there's plenty to get you started.

If you do have Python installed on your computer, you can run it from the command-line using Terminal on macOS or Windows. The traditional first program prints `Hello, world`, and the interaction looks like this:

```
$ python
Python 3.7.1 (v3.7.1:260ec2c36a, Oct 20 2018, 03:13:28)
[Clang 6.0 (clang-600.0.57)] on darwin
Type "help" [...] for more information.
>>> print("Hello, world")
Hello, world
>>>
```

Whatever you type is in **bold italic**, text printed by the computer is in the regular monospace font, and >>> is the prompt from Python itself.

If you don't have Python installed on your computer, or if you want to try an online alternative, there are a variety of services that let you run it from a web browser. Google's Colab (`colab.research.google.com`) is one of the easiest. It provides convenient access to a variety of machine-learning tools. We won't go into that here, but Colab is also good for getting started with Python. If you go to the Colab web site, select *File*, then *New notebook*, then type the program in the "+ Code" box and optionally the text in a "+ Text" box, you should see something like Figure 7.8, which shows the situation just before running the first program: a line of text that explains what the example is, and then the code itself.

Figure 7.8: Colab before running Hello world.

Clicking the triangle icon will compile and run the program; the result is shown in Figure 7.9.

The text area is used for documentation of any sort, and you can add as much code as you like. You can also add more sections of text and code as you evolve a system. Colab is a cloud version of a widely used interactive tool called a *Jupyter notebook*, a computer-based analog of a physical notebook in which you can record ideas, explanations, experiments, code and data, all in a single web page that can be edited, updated, executed, and distributed to others. More information can be found at jupyter.org.

<>   First program: hello world

    ▶   `print("Hello, world")`

    ⤷   `Hello, world`

Figure 7.9: Colab after running Hello world.

7.8 A Second Python Program

We have already seen this program, which adds up a sequence of numbers and prints the sum at the end; it's in Chapter 5. The version in Figure 7.10 prints a message along with the sum but is otherwise identical. (This program won't work if you just copy and paste it into Python, because the `input` function call is interpreted right away. You have to put it in a separate file, like `addup.py`, and then run Python on that file.)

```
sum = 0
num = input()
while num != '0':
    sum = sum + int(num)
    num = input()
print("The sum is", sum)
```

Figure 7.10: Python program to add up numbers.

Let's add it to our Colab notebook. Figure 7.11 shows the code and the state of the program just after it begins execution; the blue rectangle is where input is to be typed, the equivalent of the `prompt` dialog box that we used in JavaScript.

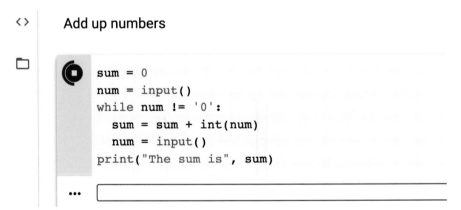

Figure 7.11: Colab before running Addup.

Figure 7.12 shows the result after I type the numbers 1, 2, 3, 4 and the 0 that terminates the loop. This version of the program includes a text message that identifies what is printed, but it does not prompt the user before each input. It's an easy and instructive exercise to add that feature.

The next example (Figure 7.13) is a program that computes the maximum value in a sequence of numbers.

```
print("The sum is", sum)
```

```
1
2
3
4
0
The sum is 10
```

Figure 7.12: Colab after running Addup.

Find maximum number

```
num = input()
max = num
while num != '0':
    if int(num) > int(max):
        max = num
    num = input()
print("The maximum is", max)
```

Figure 7.13: Colab before running Max.

Figure 7.14 shows the result after entering a sequence of numbers. Note that the program gets the right answer even when all the numbers are negative.

```
print("The maximum is", max)
```

```
-2
-5
-2
-9
0
The maximum is -2
```

Figure 7.14: Colab after running Max.

One minor change that you could make: rather than integer values, you could modify the program to use floating-point numbers, that is, numbers that potentially have fractional parts, like 3.14; the only change that is required is to replace `int` with `float` in the lines that convert input text to a numeric internal representation.

7.9 Python Libraries and Interfaces

One of Python's great strengths is the enormous collection of libraries available to Python programmers. Name an application area, and there's probably a Python library that makes it easy to write programs for that domain.

I'll illustrate with one brief example here, the `matplotlib` library for drawing graphs. Suppose we want to replicate Figure 4.1, which shows how running times grow in proportion to the amount of data. I created that figure originally in Excel, but it's easy to do the same with Python, again using our Colab notebook.

```
[ ]  import math
     import matplotlib.pyplot as plt
     log = []; linear = []; nlogn = []; quadratic = []
     for n in range(1,21):
       linear.append(n)
       log.append(math.log(n))
       nlogn.append(n * math.log(n))
       quadratic.append(n * n)
     plt.plot(linear, label="N")
     plt.plot(log, label="log N")
     plt.plot(nlogn, label="N log N")
     plt.plot(quadratic[0:10], label="N * N")
     plt.legend()
     plt.show()
```

Figure 7.15: Computing a graph of complexity classes.

The code in Figure 7.15 introduces several new features. The two `import` statements are used to access libraries of Python code, the math library and a plotting library; the latter has a long name so it's conventionally given the short alias `plt`. The values to be computed and plotted are stored in four lists, which begin life with nothing in them, indicated by the line

```
log = []; linear = []; nlogn = []; quadratic = []
```

Later statements append new values to the lists, in a loop that runs from 1 to 20 inclusive, setting the variable `n` to each value in turn. (The upper limit of the `range` is one value past the end, a Python convention that simplifies loop control.)

After the end of the loop, each list (now containing 20 items) is set up for plotting by calling the `plot` function, which will eventually draw the plot and add a label to the legend. There is one exception: only the first 10 items of the `quadratic` list are plotted, since the values grow so quickly that they would swamp the rest of the graph. The notation `[0:10]` selects a *slice* of the list containing the first ten items, which are numbered 0 through 9.

The `legend` function sets up the legend with the labels for each curve, and the `show` function generates the graph, which is shown in Figure 7.16. Matplotlib has many more features; it's worth exploring them to see how much more you can do without much effort.

Most of the program examples so far have been numeric, and it would be easy to think that programming is all about moving numbers around. But of course that's not true, as we see from all the interesting non-numeric applications in our lives.

Python's libraries make it easy to experiment with text applications as well. Figure 7.17 uses the Python `requests` library to access a copy of *Pride and Prejudice* from Gutenberg.org and print the famous opening sentence. There's a fair amount of boilerplate at the front of the book that must be skipped over. The function `find`

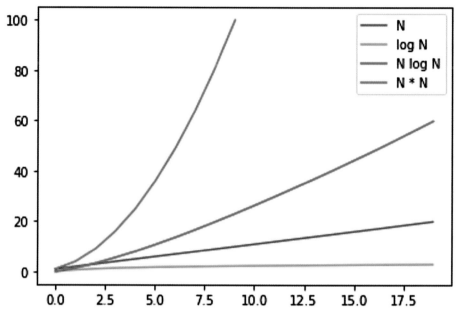

Figure 7.16: Growth of $\log N$, N, $N \log N$ and N^2.

finds the starting position of the first occurrence of some string of characters in a text, so we can use that to find the start and end locations.

The line that says

```
pandp = pandp[start:]
```

replaces the original `pandp` by the substring that begins at position `start` and goes to the end of the string.

Next, we use `find` again to locate the first period, which is at the end of the first sentence, and then print the substring that begins at position zero. Why `end+1`? The variable `end` contains the location of ".", so we need to extend it by 1 to include the period itself.

```
1 import requests
2 url = "https://www.gutenberg.org/files/1342/1342-0.txt"
3 pandp = requests.get(url).text
4 start = pandp.find("It is a truth")
5 pandp = pandp[start:]
6 end = pandp.find(".")
7 print(pandp[0:end+1])
```

⊳ It is a truth universally acknowledged, that a single man in
         possession of a good fortune, must be in want of a wife.

Figure 7.17: Accessing Internet data from Python.

In the last few examples, I've thrown a fair amount at you without much explanation, to show some of the basic ideas in a minimal form. Given this much sample code, it's easy to do simple experiments. For example, you could plot other function values like square root (sqrt) or N^3 or even 2^N. That would require changing the data ranges. You could also explore the features of matplotlib, which can do much more than we've seen here. You could download more of *Pride and Prejudice* or other texts, and explore them with Python packages for natural language processing like NLTK or spaCy.

In my experience, experimenting with existing programs is an effective way to learn more about programming, and notebooks like those enabled by Colab are a convenient way to keep track of your experiments in one place.

7.10 How Python Works

Recall the discussion of compilers, assemblers and machine instructions in Chapter 5, and the explanation of how JavaScript works, a few pages ago. A Python program is converted into executable form in an analogous way, although the details are significantly different. When you run Python, whether directly through the python command in a command-line environment or implicitly by clicking on something in a web page, the text of your program is passed to a Python compiler.

The compiler checks the program for errors, and compiles it into assembly language instructions for a made-up machine analogous to the Toy, though with a richer instruction repertoire: a virtual machine as described in Chapter 6. If there are import statements, the code from those libraries is also included. The compiler then runs a virtual machine to perform whatever actions the Python program is supposed to do. The virtual machine interacts with the environment to do operations like reading data from the keyboard or the Internet, or printing output to the screen.

If you are running Python in a command-line environment, you can use it as a high-powered calculator. You can type Python statements one at a time and each is compiled and executed right then. This makes it easy to experiment with the language and figure out what basic functions do. This is even easier when you use Python in something like a Jupyter or Colab notebook.

7.11 Summary

In the past few years, it has become trendy to encourage everyone to learn to program, with famous and influential people jumping on the bandwagon.

Should programming be a required course in elementary or high school? Should it be required in university (which my own school debates from time to time)?

My position is that it's good for anyone to know how to program. It's helpful for understanding more fully what computers do and how they do it. Programming can be a satisfying and rewarding way to spend time. The habits of thought and approaches to problem solving that programmers use transfer well to many other parts of life. And of course knowing how to program opens up opportunities; one can have a great career as a programmer and be well paid for it.

All that said, programming isn't for everyone and I don't think it makes sense to force everyone to learn to program, unlike reading, writing, and arithmetic, which are mandatory. It seems best to make the idea appealing, be sure it's easy to get started, provide plenty of opportunities, remove as many barriers as possible, and then let nature take its course.

Furthermore, computer science, often mentioned in the same discussions, is not just about programming, though programming is an important part of it. Academic computer science also involves theoretical and practical study of algorithms and data structures, which we sampled in Chapter 4. It includes architecture, languages, operating systems, networks, and an immense range of applications where computer science joins forces with other disciplines. Again, it's great for some people and many of the ideas are of broad applicability, but it's overkill to require everyone to take a formal computer science course.

Wrap-up on Software

We've covered a lot of material in the last four chapters. Here's a brief summary of the most important points.

Algorithms. An algorithm is a sequence of precise, unambiguous steps that perform some task and then stop; it describes a computation independent of implementation details. The steps are based on well-defined elementary or primitive operations. There are many algorithms; we concentrated on some of the most fundamental, like searching and sorting.

Complexity. The complexity of an algorithm is an abstract description of how much work it does, measured in terms of basic operations like examining a data item or comparing one data item with another, and expressed in terms of how that number of operations depends on the number of items. This leads to a hierarchy of complexities, ranging in our case from logarithmic at one end (doubling the number of items adds only one operation) through linear (doubling the number of items doubles the number of operations) to exponential (adding one item doubles the number of operations).

Programming. Algorithms are abstract. A program is the concrete expression of all the steps necessary to make a real computer do a complete real task. A program has to cope with limited memory and time, finite size and precision of numbers, perverse or malicious users, and a milieu of constant change.

Programming languages. A programming language is a notation for expressing all those steps, in a form that people can write comfortably but that can be translated into the binary representation that computers ultimately use. The translation can be accomplished in several ways, but in the most common case, a compiler, perhaps with an assembler, translates a program written in a language like C into binary to run on a real computer. Each different kind of processor has a different repertoire and representation of instructions, and thus needs a different compiler, though parts of that compiler may be common to different processors. An interpreter or virtual machine is a program that simulates a real or made-up computer for which code can be compiled and run; this is commonly how JavaScript and Python programs operate.

Libraries. Writing a program to run on a real computer involves a lot of intricate detail for common operations. Libraries and related mechanisms provide prefabricated components that programmers can use as they create their own programs, so that new work can build on what has already been done. Programming today is often as much gluing existing components together as it is writing original code. Components may be library functions like those we've seen in JavaScript and Python, or big systems like Google Maps and other web services. A library may be open source, so that any programmer can read, understand and improve the code, or it may be closed proprietary code. Underneath, however, they are all created by programmers writing detailed instructions in some of the languages we've talked about or others like them.

Interfaces. An interface or API is a contract between two parties: the software that provides some service and software that uses the service. Libraries and components provide their services through application programming interfaces. Operating systems make hardware look more regular and programmable through their system call interfaces.

Abstraction and virtualization. Abstraction is a fundamental idea in computing, found at all levels from hardware to large software systems. It's especially relevant in the design and implementation of software, because it separates concerns about what some piece of code does from how it is implemented. Software can be used to hide the details of implementation or to pretend to be something else; examples include virtual memory, virtual machines, interpreters, even cloud computing.

Bugs. Computers are unforgiving, and programming requires a sustained level of error-free performance from all-too-fallible programmers. Thus all big programs have bugs, and don't do exactly what they were meant to. Some bugs are mere nuisances, more like bad design than actual errors. ("That's not a bug, it's a feature" is a common saying among programmers.) Some are triggered only in such rare or unusual circumstances that they aren't even reproducible, let alone fixable. A few bugs are truly serious, with potentially grave consequences that put security, safety and even lives at stake. Liability is likely to become more of an issue in computing devices than it has been, especially as more and more critical systems are based on software. The model of "take it or leave it, without warranty" that operates in personal computing will probably be replaced by more reasonable product warranties and consumer protections as in the hardware world.

As we learn from experience, as programs are created more from proven components and as existing bugs are squashed, then in principle programs ought to become more and more free of errors. Against these advances, however, are the inevitable failures arising from continual changes as computers and languages evolve, as systems undertake new requirements, and as marketing and consumer desires create relentless pressure for new features. All of these lead to more and bigger programs. Unfortunately, bugs will always be with us.

Part III

Communications

Communications is the third major part of our four-part organization, after hardware and software. In many ways, this is where things start to get interesting (occasionally in the sense of "May you live in interesting times"), because communications involves computing devices of all types talking to each other, usually on our behalf but sometimes up to no good at all. Most technological systems now combine hardware, software, and communications, so the pieces that we have been discussing will all come together. Communications systems are also the place where most societal issues arise, presenting difficult problems of privacy, security, and the competing rights of individuals, businesses, and governments.

We'll cover a bit of historical background, talk about network technology, and then get to the Internet, which is the collection of networks that carries much of the world's computer-to-computer traffic. After that comes the (World Wide) Web, which in the mid-1990s took the Internet from a small and mostly technical user population to a ubiquitous service for everyone. We'll then turn to some of the applications that use the Internet, like mail, online commerce, and social networks, along with threats and countermeasures.

People have communicated over long distances since as far back as there are records, using much ingenuity and a remarkable variety of physical mechanisms. Every example comes with a fascinating story that would be worth a book of its own.

Long-distance runners have carried messages for millennia. In 490 BCE, Pheidippides ran 26 miles (42 kilometers) from the battlefield at Marathon to Athens, to bring news of the Athenians' great victory over the Persians. Unfortunately, at least in legend, he gasped "Rejoice, we conquer" and died.

Herodotus described the system of riders who carried messages throughout the Persian empire at about the same time; his description lives on in the 1914 inscription on the former main post office building on Eighth Avenue in New York City: "Neither snow nor rain nor heat nor gloom of night stays these couriers from the swift completion of their appointed rounds." The Pony Express, whose horseback riders carried mail 1,900 miles (3,000 km) between St. Joseph, Missouri, and Sacramento, California, is an icon of the American West, though it lasted less than two years, from

April 1860 to October 1861.

Signal lights and fires, mirrors, flags, drums, carrier pigeons, even human voices have all communicated over long distances. The word "stentorian" comes from the Greek "stentor," a person whose loud voice carried messages across narrow valleys.

One early mechanical system is not as well known as it deserves to be: the *optical telegraph* invented in France by Claude Chappe around 1792 and independently in Sweden by Abraham Edelcrantz. The optical telegraph used a signaling system based on mechanical shutters or arms mounted on towers, as seen in Figure III.1.

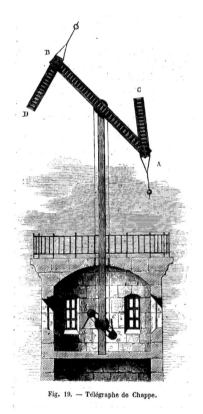

Fig. 19. — Télégraphe de Chappe.

Figure III.1: Optical telegraph station.

The telegraph operator read signals coming from the adjacent tower in one direction and passed them on to the next tower in the other direction. The arms or shutters could take on only a fixed number of positions, so the optical telegraph was truly digital. By the 1830s there was an extensive network of these towers over large parts of Europe and some parts of the United States. Towers were about 10 kilometers (6 miles) apart; transmission speeds were a few characters per minute and, according to one account, a single character could be sent from Lille to Paris (230 km or 140 miles) in about 10 minutes.

The issues that arise in modern communications systems appeared even in the 1790s. Standards were required for how information was represented, how messages were exchanged, and how errors were detected and recovered from. Sending information quickly was always a problem, though it would only take a few hours to send a short message from one end of France to the other. Security and privacy problems cropped up too. Chapter 61 of *The Count of Monte Cristo* by Alexandre Dumas, published in 1844, relates how the count bribed a telegraph operator to send a false message to Paris, causing the financial ruin of the evil banker Baron Danglars. This is a perfect example of a *man-in-the-middle* attack.

The optical telegraph had at least one major operational problem: it could only be used when visibility was good, not at night or in bad weather. The electrical telegraph, invented by Samuel F. B. Morse in the 1830s, came to fruition in the 1840s, and killed the optical telegraph within a decade. Commercial telegraph service soon linked major US cities; the first was between Baltimore and Washington in 1844 and the first trans-Atlantic telegraph cable was laid in 1858. The electrical telegraph led to many of the same hopes, aspirations and disappointments that people experienced during the early days of the Internet boom and dot-com bust of the late 1990s. Fortunes were made and lost, frauds were perpetrated, optimists predicted the coming of world peace and understanding, and realists correctly perceived that although the details were different, most of it had been seen before. "This time it's different" is rarely if ever true.

In 1876, as the story goes, Alexander Graham Bell beat Elisha Gray to the US Patent Office by a few hours with his invention, the telephone, though there is still uncertainty about the exact sequence of events. As the telephone evolved over the next hundred years, it did revolutionize communications, though it did not lead to world peace and understanding either. It let people talk to each other directly, it required no expertise to use, and standards and agreements among telephone companies made it possible to connect almost any pair of telephones in the world.

The telephone system profited from a long period of comparative stability. It only carried the human voice. A typical conversation lasted three minutes, so it didn't matter if it took a few seconds to set up the connection. A telephone number was a unique identifier with a fairly clear geographical location. The user interface was spartan, a plain black telephone with a rotary dial; these have mostly now vanished, leaving only the linguistic echo of "dialing the phone." The phone was the very antithesis of today's smartphones. All the intelligence was in the network and a user could do nothing but dial numbers to place a call, answer the phone when it rang, or ask a human operator for more complicated services. Figure III.2 shows a rotary dial phone, which was the standard for many years.

All of this meant that the telephone system could concentrate on two core values: high reliability and guaranteed quality of service. For 50 years, if one picked up the phone, there would be a dial tone (another linguistic echo), the call would always go through, the person at the other end could be heard clearly, and it would stay that way until both parties hung up. I probably have an excessively rosy view of the phone system, since I worked for over thirty years at Bell Labs, part of the American Telephone and Telegraph company (AT&T), and saw many of the changes from the inside, though far from the center of the action. On the other hand, I do miss the almost perfect reliability and clarity of pre–cell phone days.

For the phone system, the last quarter of the 20th century was a period of rapid change in technology, society and politics. The traffic model changed as fax machines became common in the 1980s. Computer-to-computer communications

Figure III.2: Rotary dial telephone (courtesy of Dimitri Karetnikov).

became common as well, using modems that converted bits into sound and vice versa; like fax machines, they used audio to pass digital data through the analog phone system. Technology made it possible to set up calls faster, to send more information (with fiber optic cables in particular, both domestically and across oceans), and to encode everything digitally. Mobile phones changed usage patterns even more dramatically; today they dominate telephony to the point where many people have abandoned wired connections at home in favor of cell phones.

Politically there was a worldwide revolution as control in the telecom industry shifted from tightly regulated companies and government agencies to deregulated private companies. This unleashed wide-open competition, which resulted in a downward spiral in revenue for the old-line telephone companies and the rise and often fall of a host of new players.

Today, telecom companies continue to wrestle with the threats raised by new communications systems, primarily those based on the Internet, often in the face of declining revenue and market share. One threat comes from Internet telephony. It's easy to send digital voice over the Internet. Services like Skype push this further, offering free computer-to-computer voice and video and a way to call from the Internet to conventional phones for a nominal price, usually much less than the incumbent phone companies charge, especially for international calling. The handwriting was on the wall long ago, though not everyone saw it. I recall a colleague telling AT&T management in the early 1990s that prices for domestic long distance phone calls would drop to a cent a minute—at a time when AT&T was charging more like 10 cents a minute—and he was laughed at.

Similarly, cable companies like Comcast are threatened by streaming services provided by Netflix, Amazon, Google, and many others, all of which use the Internet, reducing the cable company to merely carrying someone else's bits.

Naturally, incumbent companies are fighting to retain their revenues and effective monopolies by technical, legal and political means. One approach is to charge competitors for access to the wired phones in residences. Another approach is to put bandwidth limitations and other slowdowns in the path of competitors that provide telephone service using the Internet ("voice over IP" or "VoIP") or other services.

This is related to a general issue called *net neutrality*. Should Internet service providers be allowed to interfere with, degrade, or block traffic for any reason other than purely technical ones related to efficient network management? Should telephone and cable companies be required to provide the same level of Internet service to all users, or should they be able to treat services and users differently? If so, on what grounds? For instance, should a telephone company be allowed to slow down traffic from a competitor, say a VoIP company like Vonage? Should a cable and entertainment company like Comcast be allowed to slow down traffic for Internet movie services like Netflix that it competes with? Should service providers be allowed to impede traffic for sites that espouse social or political views that the owners disagree with? As usual, there are arguments on both sides.

The resolution of the net neutrality issue will have a significant effect on the future of the Internet. So far the Internet has generally provided a neutral platform that carries all traffic without interference or restrictions. This has benefited everyone and in my opinion, it's highly desirable to maintain this state of affairs.

On the other hand, the Internet supports a wide variety of sites that provide forums for misinformation, fake news, bigotry and mysogyny, hate speech, conspiracy theories, libel, and any number of other undesirable activities. In the US at least, there is an ongoing discussion over whether Internet sites like Twitter and Facebook are merely platforms for communication, and thus not responsible for the content that they host, just as phone companies are not responsible for what people say when they use the telephone. Or are they publishers like newspapers who must take some responsibility for what is published on their sites? Not surprisingly, positions taken depend on what problem is being avoided, but for the most part, social media sites do not want to be considered publishers.

8

Networks

"Mr. Watson—Come here—I want to see you."

First intelligible message sent by telephone,
Alexander Graham Bell, March 10, 1876.

In this chapter, I'm going to talk about network technologies that one encounters directly in daily life: conventional wired networks like telephones, cable, and Ethernet, and then wireless networks, of which Wi-Fi and cell phones are the most common. These are how most people connect to the Internet, which is the topic of Chapter 9.

All communications systems share basic properties. At the source, they convert information into a representation that can be transmitted over some medium. At the destination, they convert that representation back into a usable form.

Bandwidth is the most fundamental property of any network—how fast the network can transmit data. This ranges from a few bits per second for systems that operate under severe power or environmental constraints to terabits per second for the fiber optic networks that carry Internet traffic across continents and oceans. For most people, bandwidth is the property that matters the most. If there's enough bandwidth, data flows quickly and smoothly; if not, communication is a frustrating experience of halting and stuttering.

Latency or *delay* measures how long it takes for a particular chunk of information to go through the system. High latency need not mean low bandwidth: driving a truck full of disk drives across the country has high delay but enormous bandwidth.

Jitter—the variability of delay—also matters in some communications systems, especially those dealing with speech and video.

Range defines how geographically big a network can be with a given technology. Some networks are local, a few meters at most, while others literally span the world.

Other properties include whether the network broadcasts so that multiple receivers can hear one sender (as with radio), or is point to point, pairing a specific sender and receiver. Broadcast networks are intrinsically more vulnerable to eavesdropping, which may have security implications. One has to worry about what kinds of errors can occur and how they are handled. Other factors to consider include the cost of hardware and infrastructure, and the amount of data to be sent.

8.1 Telephones and Modems

The telephone network is a large and successful worldwide network that began by carrying voice traffic and eventually evolved to carrying considerable data traffic as well. In the early days of home computers, most users were connected online by phone lines.

At the residential level, the wired telephone system still carries mostly analog voice signals, not data. So to send digital data, it's necessary to have a device that converts bits to sound and back again. The process of imposing an information-carrying pattern on a signal is called *modulation*. At the other end, it's necessary to convert the pattern back into its original form, which is called *demodulation*. The device that does modulation and demodulation is called a *modem*. A telephone modem used to be a large and expensive separate box of electronics, but today it's a single chip and is practically free. In spite of that, the use of wired telephones to connect to the Internet is uncommon now, and few computers have modems.

Using the telephone for data connections has major drawbacks. It requires a dedicated phone line, so if you only have one phone line in your home, you have to choose between connecting or leaving the phone available for voice calls. More important for most people, however, is that there is a tight limit on how quickly information can be sent by telephone. The maximum speed is about 56 Kbps (56,000 *bits* per second—a lower case "b" conventionally stands for bits, in contrast to the upper case "B" that stands for bytes), which is 7 KB per second. A 20 KB web page thus takes 3 seconds to download, a 400 KB image takes nearly 60 seconds, and a video or a software update could take hours or even days.

8.2 Cable and DSL

The 56 Kbps limitation on how fast an analog telephone line can carry signals is inherent in its design, an artifact of engineering decisions made 60 years ago, at the beginning of the transition to a digital telephone system. Two other technologies provide an alternative for many people, with at least 100 times the bandwidth.

The first is to use the cable that carries cable television into many homes. That cable can carry hundreds of video channels simultaneously. It has enough excess capacity that it can be used to carry data to and from homes as well; a cable system will offer a wide range of download speeds (and prices), typically a few hundred Mbps. The device that converts signals from the cable into bits for a computer and back again is called a *cable modem*, since it does modulation and demodulation just like a telephone modem, though it runs quite a bit faster.

The high speed is illusory in a way. The same TV signal goes to every house, regardless of whether it's being watched or not. On the other hand, although the cable is a shared medium, data going to my house is meant for me and won't be the same data at the same time as is going to your house, so there's no way for us to share the content. Data bandwidth has to be shared among data users of the cable, and if I'm using a lot of it, you won't get as much. More likely, we'll both get less. Fortunately, we're not likely to interfere with each other too much. It's rather like a communications version of deliberate over-booking by airlines and hotels. They know that not everyone will show up, so they can safely over-commit resources. It works in communications systems too.

Now you can see another problem. We all watch potentially the same TV signals, but I don't want my data going to your house any more than you want your data going to my house. After all, it's personal—it includes my email, my online shopping and banking information, and perhaps even personal entertainment tastes that I would rather not have others know about. This can be addressed by encryption, which prevents anyone else from reading my data; we'll talk more about this in Chapter 13.

There is yet another complication. The first cable networks were one-way: signals were broadcast to all homes, which was easy to build but provided no way to send information back from the customer to the cable company. Cable companies had to find a way to deal with this anyway, to enable pay-per-view and other services that require communication from the customer. Thus cable systems became two-way, which makes them usable as communications systems for computer data. It is common, however, for the upload speed (from consumer to cable company) to be much lower than the download speed, since most of the traffic is downloading.

The other reasonably fast network technology for home use is based on the other system that's already in the home, the good old telephone. It's called Digital Subscriber Loop or *DSL* (sometimes *ADSL*, for "asymmetric," because the bandwidth down to the home is higher than the bandwidth up from the home). It provides much the same service as cable, but with major differences underneath.

DSL sends data on the telephone wire with a technique that doesn't interfere with the voice signal, so you can talk on the phone while surfing, and neither affects the other. This works well but only up to a certain distance. If you live within about 3 miles (5 km) of a local telephone company switching office, as many people do, you can have DSL, but if you're too far away, you're out of luck.

The other nice thing about DSL is that it's not a shared medium. It uses the dedicated wire between your home and the phone company, but no one else does, so you don't share the capacity with your neighbors, nor do your bits go to their homes. A special box at your house—another modem, with a matching one at the telephone company's building—converts signals into the right form to be sent along the wires. Otherwise, cable and DSL look and feel much the same. The prices tend to be about the same too, at least when there is competition. Anecdotally, however, DSL use in the US seems to be decreasing.

Technology continues to improve, and home fiber optic service is replacing the older coaxial cable or copper wires. For example, Verizon recently replaced their aging copper connection to my home with optical fiber, which is cheaper to maintain, and allows them to offer additional services like Internet access. The only downside from my standpoint (aside from a few days without service because they inadvertently cut a cable during the installation) is that if there's an extended power failure I won't have phone service. In olden times, phones got their power from batteries and generators at the phone company's facility, and would work in spite of power failures; that's not true of fiber optic cables.

Fiber optic systems are much faster than the alternatives. Signals are sent as pulses of light along an extremely pure glass fiber with low loss; signals can propagate for kilometers before they need to be amplified back to full strength. In the early 1990s, I was part of a "fiber to the home" research experiment and for a decade had a 160 Mbps connection to my house. That gave me serious bragging rights but not much else, since there wasn't any service that could take advantage of that much bandwidth.

Today, through another accident of geography, I have a gigabit fiber connection into my home (not the one from Verizon), but the effective rate is only 30 to 40 Mbps because it's limited by my home wireless router. On my office wireless network, I get about 80 Mbps to a laptop, but 500 to 700 Mbps to a computer on an Ethernet connection. You can test your own connections at sites like `speedtest.net`.

8.3 Local Area Networks and Ethernet

Telephone and cable are network technologies that connect a computer to a larger system, usually at a considerable distance. Historically, there was another thread of development that led to one of today's most common network technologies, Ethernet.

In the early 1970s, Xerox's Palo Alto Research Center (Xerox PARC) developed an innovative computer called the Alto that served as a vehicle for experiments that led to numerous other innovations. It had the first window system and a bitmap display that was not restricted to displaying characters. Although Altos were too expensive to be personal computers in today's sense, every researcher at PARC had one.

One problem was how to connect Altos to each other or to a shared resource like a printer. The solution, invented by Bob Metcalfe and David Boggs in the early 1970s, was a networking technology that they called *Ethernet*. An Ethernet carried signals between computers that were all connected to a single coaxial cable, physically similar to the one that carries cable TV into your house today. The signals were pulses of voltage whose strength or polarity encoded bit values; the simplest form might use a positive voltage for a 1-bit and a negative voltage for a 0-bit.

Each computer was connected to the Ethernet by a device with a unique identification number. When one computer wanted to send a message to another, it listened to make sure no one else was already sending, then broadcast its message onto the cable along with the identification number of the intended recipient. Every computer on the cable could hear the message, but only the computer to which it was directed would read and process it.

Every Ethernet device has a 48-bit identification number, different from all other devices, called its (Ethernet) address. This allows for 2^{48} (about 2.8×10^{14}) devices in all. You can find the Ethernet address for your computer, since it is sometimes printed on the bottom of the computer and it can also be displayed by programs like `ipconfig` on Windows or `ifconfig` on Macs, or found in System Preferences or Settings. Ethernet addresses are always written in hexadecimal with two digits per byte, so there are 12 hex digits in all. Look for some sequence of hexadecimal digits like `00:09:6B:D0:E7:05` (with or without the colons), though since that's from one of my laptops, you won't likely find exactly that one on your computer.

From the discussion of cable systems above, you can imagine that Ethernets will have similar problems of privacy and contention for a limited resource.

Contention is handled by a neat trick: if a network interface starts to send but detects that someone else is sending too, it stops, waits a brief period, and tries again. If the waiting time is random and gradually increases over a series of failures, then eventually everything goes through. It's sort of like conversation at a party: if two people start to say something at the same time, both back off, then one restarts before the other.

Privacy wasn't originally a concern since everyone was an employee of the same company and all were working in the same small building. Privacy is a major problem today, however. It's possible for software to put an Ethernet interface into "promiscuous mode," in which it reads the contents of all messages on the network, not just those intended for it specifically. That means it can look for interesting content like unencrypted passwords. Such "sniffing" used to be a common security problem on Ethernet networks in college dorms. Encryption of packets on the cable is a solution, and today most traffic is encrypted by default.

You can experiment with sniffing with an open-source program called Wireshark, which displays information about Ethernet traffic, including wireless. I occasionally demonstrate Wireshark in class when it seems that students are paying more attention to their laptops and phones than to me; the demo does catch their attention, albeit briefly.

Information on an Ethernet is transmitted in packets. A *packet* is a sequence of bits or bytes that contains information in a precisely defined format so that it can be packed up for sending and cracked open when received. If you think of a packet as an envelope (or perhaps a postcard) with the sender's address, the recipient's address, the content, and miscellaneous other information, in a standard format, that's a reasonably good metaphor, as are the standardized packages used by shipping companies like FedEx.

The details of packet format and content vary greatly among networks. An Ethernet packet (Figure 8.1) has six-byte source and destination addresses, some miscellaneous information, and up to about 1500 bytes of data.

source address	destination address	data length	data (48-1518 bytes)	error check

Figure 8.1: Ethernet packet format.

Ethernet has been an exceptionally successful technology. It was first made into a commercial product (not by Xerox, but by 3Com, a company founded by Metcalfe), and over the years billions of Ethernet devices have been sold by a large number of vendors. The first version ran at 3 Mbps, but today's versions are anywhere from 100 Mbps to 10 Gbps. As with modems, the first devices were bulky and expensive, but today an Ethernet interface is a single inexpensive chip.

Ethernets have a limited range, a few hundred meters. The original coaxial cable has been superseded by an 8-wire cable with a standard connector that lets each device plug in to a "switch" or "hub" that broadcasts incoming data to the other connected devices. Desktop computers usually have a socket that accepts this standard connector, which also appears on devices like wireless base stations and cable modems that simulate Ethernet behavior; the socket has disappeared from modern laptops, which rely on wireless networking.

8.4 Wireless

Ethernet has one significant drawback: it needs wires, real physical equipment that snakes through walls, under floors, and sometimes (I speak from personal experience here) across the hall, down the stairs, and through the dining room and kitchen on its way to the family room. A computer on an Ethernet can't be easily moved around, and if you like to lean back with your laptop on your lap, an Ethernet cable is a nuisance.

Fortunately, there's a way to have your cake and eat it too, with wireless. A wireless system uses radio to carry data, so it can communicate from any place where there is enough signal. The normal range for wireless networks is tens to hundreds of meters. Unlike infrared, which is used for TV remotes, wireless doesn't have to be line of sight because radio waves can pass through some materials, though not all. Metal walls and concrete floors interfere with radio waves, so in practice the range can be less than it would be in open air. Higher frequencies are generally absorbed more than lower frequencies, all else being equal.

Wireless systems use electromagnetic radiation to carry signals. The radiation is a wave of a particular frequency measured in Hz or more likely MHz or GHz for the systems we encounter, like the 103.7 MHz of a radio station. A modulation process imposes an information signal onto the carrier wave. For example, amplitude modulation ("AM") changes the amplitude or strength of the carrier to convey information, while frequency modulation ("FM") changes the frequency of the carrier around its central value. The received signal strength varies directly with power level at the transmitter and inversely as the square of the distance from the transmitter to the receiver; thus a receiver twice as far away as another will receive a signal only one quarter as strong.

Wireless systems operate under strict rules about what range of frequencies they can use—their *spectrum*—and how much power they can use for transmission. Spectrum allocation is a contentious process, since there are many competing demands on it. Spectrum is allocated by government agencies like the Federal Communications Commission (FCC) in the United States, and international agreements are coordinated by the International Telecommunication Union or ITU, which is an agency of the United Nations. In the US, when new spectrum space becomes available, most often of very high frequency bands, it is typically allocated by public auctions run by the FCC.

The wireless standard for computers has the catchy name IEEE 802.11, though you will more often see the term *Wi-Fi*, which is a trademark of the Wi-Fi Alliance, an industry group. IEEE is the Institute of Electrical and Electronics Engineers, a professional society that among other activities establishes standards for a wide variety of electronic systems, including wireless. 802.11 is the number of the standard, which has more than a dozen parts for different speeds and underlying technologies. Nominal speeds range up to nearly a gigabit per second, but these rates overstate the speed that one might achieve in real-world conditions.

A wireless device encodes digital data into a form suitable for carrying on radio waves. A typical 802.11 system is packaged so that it behaves like an Ethernet. The range is likely to be similar, but there are no wires to contend with.

Wireless Ethernet devices operate at frequencies around 2.4–2.5 GHz, 5 GHz, and higher frequencies as well. When wireless devices all use the same narrow frequency band, contention is a distinct possibility. Worse, other devices use this same overcrowded band, including some cordless telephones, medical equipment, and even microwave ovens.

I'm going to briefly describe three wireless systems that are in widespread use. The first is Bluetooth, which is named after the Danish king Harald Bluetooth (c. 935–985). Bluetooth is intended for short-range ad hoc communications. It uses the same 2.4 GHz frequency band as 802.11 wireless. The range is from 1 to 100 meters, depending on power level, and the data rate is 1 to 3 Mbps. Bluetooth is used in TV remote controls, wireless microphones, ear buds, keyboards, mice, and game controllers, where low power consumption is critical; it's also used in cars for hands-free use of phones.

RFID or radio-frequency identification is a low-power wireless technology used in electronic door locks, identification tags for a variety of goods, automatic toll systems, implanted chips in pets, and even in documents like passports. The tag is basically a small radio receiver and transmitter that broadcasts its identification as a stream of bits. Passive tags do not have batteries, but obtain power from an antenna that receives a signal broadcast by an RFID sensor; when the chip is near enough to a sensor, typically only a few inches, it responds with its identifying information. RFID systems use a variety of frequencies, though 13.56 MHz is typical. RFID chips make it possible to quietly monitor where things and people are. Chip implants in pets are popular—our cat has one so she can be identified if she gets lost—and as might be expected, there have been suggestions for implanting people as well, for both good and bad reasons.

The Global Positioning System (GPS) is an important one-way wireless system, commonly seen in car and phone navigation systems. GPS satellites broadcast precise time and location information, and a GPS receiver uses the time it takes for signals to arrive from three or four satellites to compute its position on the ground. But there is no return path. It's a common misconception that GPS somehow tracks its users. As a story in the *New York Times* had it a few years ago, "Some [cell phones] rely on a global positioning system, or GPS, which sends the signal to satellites that can pinpoint almost exactly where a user is." This is wrong. GPS-based tracking needs ground-based systems like cell phones to relay the location. A cell phone is in constant communication with base stations, as we will discuss next, so the phone company knows your precise location whenever your phone is on. When you enable location services, that information is also made available to apps.

8.5 Cell Phones

The most common wireless communication system for most people is the cell or mobile phone, now usually just "cell" or "mobile," a technology that barely existed in the 1980s, but which is now used by well over half of the world's population. Cell phones are a case study for the kinds of topics that are covered in this book—interesting hardware, software and of course communications, with plenty of social, economic, political and legal issues to go along with them.

The first commercial cell phone system was developed by AT&T in the early 1980s. The phones were heavy and bulky; advertisements of the time show users carrying a small suitcase for the batteries while standing next to a car that carries the antenna.

Why "cell"? Because spectrum and radio range are both limited, a geographical area is divided into "cells," rather like imaginary hexagons (Figure 8.2), with a *base station* in each cell; the base station is connected to the rest of the telephone system. Phones talk to the closest base station, and when they move from one cell to another, a call in progress is handed off from the old base station to the new one; most of the time the user doesn't know this has happened.

Since received power falls off in proportion to the square of the distance, frequency bands within the allotted spectrum can be re-used in non-adjacent cells without significant interference; this was the insight that made it possible to use the limited spectrum effectively. In the diagram of Figure 8.2, the base station at 1 will not share frequencies with base stations 2 through 7, but can share with 8 through 19 because they are far enough away to avoid interference. The details depend on factors like antenna patterns; the diagram is an idealization.

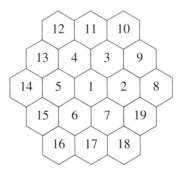

Figure 8.2: Cell phone cells.

Cell sizes vary, from a few hundred meters to a few tens of kilometers in diameter, depending on traffic, terrain, obstacles, and the like.

Cell phones are part of the regular telephone network, but are connected to it by radio via the base stations instead of wires. The essence of cell phones is mobility. Phones move long distances, often at high speed, and can appear in a new location with no warning at all, as they do when turned on again after a long flight.

Cell phones share a narrow radio frequency spectrum that has limited capacity for carrying information. Phones must operate with low radio power because they use batteries, and by law their transmission power is limited to avoid interference with others. The bigger the battery, the longer the time between charges, but also the bigger and heavier the phone; this is another of the tradeoffs that designers must make.

Cell phone systems use different frequency bands in different parts of the world, but are generally around 900 and 1900 MHz; newer phone standards like 5G also use much higher frequencies. Each frequency band is divided into multiple channels, and a conversation uses one channel for each direction. Signaling channels are shared by all phones in the cell; they are also used for text messages and data in some systems.

Each phone has a unique 15-digit identification number called its International Mobile Equipment Identity, or IMEI, analogous to an Ethernet address. When the phone is turned on, it broadcasts its identification number. The nearest base station hears the phone and validates it with the home system. As the phone moves around, its location is kept up to date by base stations reporting to the home system; when someone calls the phone, the home system knows which base station is currently in contact with the phone.

Phones talk to the base station with the strongest signal. The phone continuously adjusts its power level, to use less power when it's closer to the base station; this preserves its own battery and creates less interference for other phones. Merely keeping in touch with a base station uses much less power than a phone call does, which is why standby times are measured in days while talk times are in hours. If the phone is in an area of weak or non-existent signal, however, it will use up its battery faster as it looks in vain for a base station.

All phones use data compression to squeeze the signal into as few bits as possible, then add error correction to cope with the inevitable errors of sending data over a noisy radio channel in the face of interference. We'll come back to these shortly.

Mobile phones raise political and social issues. Spectrum allocation is clearly one; in the US, the government restricts the use of the allotted frequencies to at most two companies in each band. Thus spectrum is a valuable resource. One of the driving forces behind the merger of Sprint and T-Mobile in 2020 was to make better use of their somewhat separate spectrum holdings.

Cell tower locations are another source of potential conflicts. Cell phone towers are not the most esthetic of outdoor structures; for instance, Figure 8.3 shows a "Frankenpine," a cell tower imperfectly disguised as a tree. Many communities don't want such towers within their borders, though of course they want high quality phone service.

Cell phone traffic is vulnerable to a targeted attack by a device known generically as a *stingray*, after a commercial product called "StingRay." A stingray mimics a cell tower so that nearby cell phones communicate with the device rather than a real tower. This can be used for passive surveillance or active engagement with the cell phone (a man-in-the-middle attack). Phones are designed to communicate with the base station that provides the strongest signal; a stingray thus works in a small area where it can present a stronger signal than any nearby cell tower.

Local law enforcement agencies in the US appear to be using stingray devices in growing numbers, but have been trying to keep their use secret or at least low profile. It's not at all clear that their use to collect information about potential criminal activity is legal.

Figure 8.3: Cell phone tower disguised as a tree.

On the social level, cell phones have revolutionized many aspects of life. We use smartphones less for conversation than for all their other features. Phones have become the primary form of Internet access, since they provide browsing, mail, shopping, entertainment and social networking, albeit on a small screen. Indeed, there is some convergence between laptops and cell phones as the latter become more powerful while remaining highly portable. Phones have also taken over the functions of other devices, from watches and address books to cameras, GPS navigators, fitness trackers, voice recorders, and music and movie players.

Downloading movies to a phone requires a lot of bandwidth. As cell phone use expands, the strain on existing facilities is only going to increase. In the US, carriers apply usage-sensitive pricing and bandwidth caps to their data plans, ostensibly to restrain bandwidth hogs who download full-length movies, though the caps apply even when traffic is light.

It's also possible to use a mobile phone as a *hotspot* that lets you connect your computer to the Internet through the phone's cellular connection. This is sometimes called "tethering." Carriers may apply limits and extra charges, since a hotspot can also use a lot of bandwidth.

8.6 Bandwidth

Data on a network flows no faster than the slowest link permits. Traffic can slow down at many places, so bottlenecks are common on the links themselves and in the processing at computers along the way. The speed of light is also a limiting factor. Signals propagate at 300 million meters per second in vacuum (about one foot per nanosecond, as Grace Hopper so often observed) and more slowly in electronic circuits, so it takes time to get a signal from one place to another even if there are no other delays. At the speed of light in a vacuum, the time to travel from the east coast of the US to the west coast (2,500 miles or 4,000 km) is about 13 milliseconds. For comparison, a typical Internet delay for the same trip is about 40 msec, Paris is about 50 msec, Sydney is 110 msec and Beijing is 140 msec. The times are not necessarily in order of physical distance.

We encounter a wide range of bandwidths in day-to-day life. My first modem ran at 110 bits per second or bps, which was fast enough to keep up with a mechanical typewriter-like device. Home wireless systems running 802.11 can in theory operate up to 600 Mbps, though in practice the rates will be much lower. Wired Ethernets are commonly 1 Gbps. A cable connection between your home and your Internet service provider might be a few hundred megabits per second if it uses optical fiber. Your ISP is likely connected to the rest of the Internet through fiber optic links that could in principle provide 100 Gbps or more.

Phone technology is exceptionally complicated, and changes continuously in the quest for higher bandwidth. Cell phones operate in such complex environments that it's hard to assess their effective bandwidth. Most phones today use a standard called *4G*, or fourth generation, and the industry is moving to the next generation, not surprisingly called *5G*. 3G phones still exist but appear to be an endangered species in the US; a recent message from my carrier warns that within a year one of my phones will no longer work.

4G phones are supposed to provide about 100 Mbps for moving environments like cars and trains, and 1 Gbps for stationary or slowly moving phones. These speeds seem to be more aspirational than real, and there is plenty of room for optimistic advertising. That said, my 4G phone is fast enough for my low-intensity uses, like mail, occasional browsing and interactive maps.

You will sometimes see the term *4G LTE*. *LTE* stands for Long-Term Evolution, which is not a standard but a sort of road map for the path from 3G to 4G. Phones that are somewhere on that path may display "4G LTE" to show that they are at least heading towards 4G.

The first deployments of 5G began in 2019. Phones that use the 5G standard will have higher bandwidth, at least when connected to the right equipment at the right distance; the nominal speed range is from 50 Mbps to as much as 10 Gbps. The phones use up to three frequency ranges, the lower two of which are used by existing 4G phones as well, so 5G is similar to 4G in these bands. 5G uses much higher frequencies for short-range connections (roughly 100 meters), and these allow for higher speeds. 5G also allows many more devices in a given area, which will help as Internet of Things devices begin to use 5G.

8.7 Compression

One way to make better use of available memory and bandwidth is to compress data. The basic idea of compression is to avoid storing or sending redundant information, that is, information that can be recreated or inferred when retrieved or received at the other end of a communications link. The goal is to encode the same information in fewer bits. Some bits carry no information and can be removed entirely; some bits can be computed from others; some don't matter to the recipient and can be safely discarded.

Consider English text like this book. In English, letters do not occur with equal frequency; "e" is most common, followed by "t," "a," "o," "i" and "n" in approximately that order; at the other end, "z," "x" and "q" are much less common. In the ASCII representation of text, each letter occupies one byte or 8 bits. One way to save a bit (so to speak) is to use only 7 bits; the 8th bit (that is, the leftmost) is always zero in US ASCII and thus carries no information.

We can do even better by using fewer bits to represent the most common letters and if necessary more bits to represent infrequent letters, reducing the total bit count significantly. This is analogous to the approach taken by Morse code, which encodes the frequent letter "e" as a single dot, "t" as a single dash, but the infrequent "q" as dash-dash-dot-dash.

Let's make this more concrete. *Pride and Prejudice* has somewhat over 121,000 words, or 680,000 bytes. The most common character is the space between words: there are nearly 110,000 spaces. The next most common characters are e (68,600), t (456,900), and a (31,200); at the other end, Z occurs only three times, and X not at all. The least common lower case letters are j 551 times, q 627 times, and x at 839. Clearly if we used two bits each for space, e, t, and a, we'd save a great deal and it wouldn't matter if we had to use more than 8 bits for X, Z, and other infrequent letters. An algorithm called Huffman coding does this systematically, finding the best possible compression that encodes individual letters; it compresses *Pride and Prejudice* by 44 percent, to 390,000 bytes, so the average letter requires about four and a half bits.

It's possible to do even better by compressing larger chunks than single letters, for example entire words or phrases, and by adapting to the properties of the source document. Several algorithms do this well. The widely used ZIP compression algorithm squeezes the book by 64 percent, down to 249,000 bytes. The Unix program called bzip2 gets it down to 175,000 bytes, barely one quarter of its original size.

Images can also be compressed. Two common forms are GIF (Graphics Interchange Format) and PNG (Portable Network Graphics), which are intended for images that are primarily text, line art and blocks of solid colors. GIF supports only 256 distinct colors, but PNG supports at least 16 million. Neither is intended for photographic images.

All of these techniques do *lossless compression*—the compression loses no information, so uncompressing restores the original source exactly. Although it might seem counterintuitive, there are situations where it's not necessary to reproduce the original input exactly—an approximate version is good enough—and in those situations, a *lossy* compression technique can give even better results.

Lossy compression is most often used for content that is going to be seen or heard by people. Consider compressing an image from a digital camera. The human eye can't distinguish colors that are close to each other, so it's not necessary to preserve the exact colors of the input; fewer colors will do and those can be encoded in fewer bits. Similarly it's possible to discard some fine detail; the resulting image will not be as sharp as the original but the eye won't notice. The same is true for fine gradations of brightness. The JPEG compression algorithm that produces the ubiquitous .jpg images uses this to compress a typical image by a factor of 10 or more without significant degradation as seen by our eyes. Most programs that produce JPEG allow some control over the amount of compression; "higher quality" means that there is less compression.

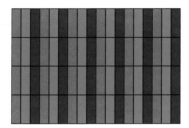

Figure 8.4:   RGB pixels.

Figure 2.2, reproduced in Figure 8.4, is the kind of image that PNG compression is meant for. At its original size of about 2 inches (5 cm) wide, it occupies about 10 KB. A JPEG version is 25 KB, and when viewed close up, shows obvious visual artifacts that are not in the original. On the other hand, photographs will compress better with JPEG.

The MPEG family of algorithms for compressing movies and TV is also a perceptual technique. Individual frames can be compressed as in JPEG but in addition, it's possible to compress a sequence of blocks that do not change much from one frame to the next. It's also possible to predict the result of motion and encode only the change, and even to separate a moving foreground from a static background, using fewer bits for the latter.

MP3 and its successor AAC are the audio part of MPEG; they are *perceptual coding* algorithms for compressing sound. Among other things, they take advantage of the fact that louder sounds mask softer sounds, and that the human ear can't hear frequencies higher than about 20 KHz, a number that falls off as one gets older. The encoding generally compresses standard CD audio by a factor of about 10.

Cell phones use a great deal of compression. Voice can be squeezed significantly more than arbitrary sound can be, because it has a narrow range of frequencies and it's produced by a vocal tract that can be modeled for individual speakers; using the characteristics of an individual person makes better compression possible.

The idea in all forms of compression is to reduce or eliminate bits that do not convey their full potential information content, by encoding more frequently occurring elements in fewer bits, by building dictionaries of frequent sequences, and by encoding repetitions with counts. Lossless compression allows the original to be reconstructed perfectly; lossy compression discards some information that is not needed by the recipient, and offers a tradeoff between quality and compression factor.

It's possible to make other tradeoffs as well, for example between compression speed and complexity versus decompression speed and complexity. When a digital television picture breaks up into blocks or the sound starts to become garbled, that's a result of the decompression algorithm failing to recover from some input error, probably because the data didn't arrive fast enough. Finally, no matter what the algorithm, some inputs will not shrink, as you can see by imagining repeatedly applying the algorithm to its own output; in fact, some inputs will become larger.

It's hard to imagine, but compression can even be the stuff of entertainment. The HBO television series *Silicon Valley*, which premiered in 2014 and ran for 53 episodes over 6 seasons, is based on the invention of a novel compression algorithm and the struggles of its inventor to protect his startup company against larger companies that want to steal his ideas.

8.8 Error Detection and Correction

If compression is the process of removing redundant information, error detection and correction is the process of adding carefully controlled redundancy that makes it possible to detect and even correct errors.

Some common numbers have no redundancy and thus it's not possible to detect when an error might have occurred. For example, US Social Security numbers have 9 digits and almost any 9 digit sequence could be a legitimate number. (This is helpful when someone asks for yours when they don't need it; just make one up.) But if some extra digits were added or if some possible values were excluded, it would be possible to detect errors.

Credit card and cash machine numbers are 16 digits long but not every 16-digit number is a valid card number. They use a checksum algorithm, invented by Hans Peter Luhn at IBM in 1954, that detects single-digit errors and most transposition errors in which two digits are interchanged; those are the most common kinds of errors that occur in practice.

The algorithm is easy: starting at the rightmost digit, multiply each successive digit alternately by 1 or by 2. If the result is greater than 9, subtract 9. Add the resulting digits; the sum must be divisible by 10. Test it on your own cards and on 4417 1234 5678 9112, a number that some banks use in advertisements. The result on the latter is 9, so this is not a valid number, but changing the final digit to 3 makes it valid.

The 10- or 13-digit ISBNs on books also have a checksum that defends against the same kinds of errors, using a similar algorithm.

Those algorithms are special-purpose and aimed at decimal numbers. A *parity code* is the simplest example of general purpose error detection that is applied to bits. A single additional *parity bit* is attached to each group of bits; the parity bit's value is chosen so that the total number of (say) 1-bits in the group is even. That way if a single bit error occurs, the receiver will see an odd number of 1-bits and know that something has been damaged. Of course this doesn't identify which bit was in error, nor does it detect the occurrence of two errors.

For example, Figure 8.5 shows the first half-dozen upper case letters in ASCII, written in binary. The even-parity column replaces the unused leftmost bit by a parity bit that makes the parity even (each byte has an even number of 1-bits), while in the odd-parity column each byte has an odd number of 1-bits. If any bit in any of these is flipped, the resulting byte does not have the correct parity and thus the error can be detected. If a few more bits were used, a code could correct single bit errors.

letter	original	even parity	odd parity
A	01000001	01000001	11000001
B	01000010	01000010	11000010
C	01000011	11000011	01000011
D	01000100	01000100	11000100
E	01000101	11000101	01000101
F	01000110	11000110	01000110
. . .			

Figure 8.5: ASCII characters with even and odd parity bits.

Error detection and correction are used extensively in computing and communications. Error correction codes can be used on arbitrary binary data, though different algorithms are chosen for different kinds of likely errors. For example, some primary memories use parity bits to detect single-bit errors at random places; CDs and DVDs use codes that can correct long runs of damaged bits; cell phones can cope with short bursts of noise. QR codes like the one in Figure 8.6 are two-dimensional barcodes with a lot of error correction. As with compression, error detection can't solve all problems, and some errors will always be too severe to be detected or corrected.

Figure 8.6: QR code for http://www.kernighan.com.

8.9 Summary

Spectrum is a critical resource for wireless systems, and there is never enough for the demand. Many parties compete for spectrum space, with established interests like broadcasting and telephone companies resisting change. One way to cope is to use existing spectrum more efficiently. Cell phones originally used analog encoding but those systems were phased out long ago in favor of digital systems that use much less bandwidth. Sometimes existing spectrum is repurposed; the switch to digital TV in the US in 2009 freed up a large block of spectrum space for other services to fight over. Finally, it's possible to use higher frequencies, but that usually means shorter range; effective range decreases with the square of the frequency, another example of a quadratic effect.

Wireless is a broadcast medium, so anyone can listen in; encryption is the only way to control access and protect information in transit. The original standard for wireless encryption on 802.11 networks (WEP, or Wired Equivalent Privacy) proved to have significant weaknesses; current encryption standards like WPA (Wi-Fi Protected Access) are better. Some people still run open networks, that is, with no encryption at all, in which case anyone nearby can not only listen in but can use the wireless service for free. Anecdotally, the number of open networks is much lower today than a few years ago, as people have become more sensitive to the perils of eavesdropping and free riding.

Free Wi-Fi services in coffee shops, hotels, airports, and so on are an exception; for example, coffee shops want their customers to linger (and buy expensive coffee) while using their laptops. Information passing over those networks is open to all unless you use encryption, and not all servers will encrypt on demand. Furthermore, not all open wireless access points are legitimate; sometimes they are set up with the explicit intent of trapping naive users. You should not do anything sensitive over a public network and be especially careful about using access points that you don't know anything about.

Wired connections will always be a major network component behind the scenes, especially for high bandwidth and long distances. Nevertheless, in spite of limits on spectrum and bandwidth, wireless will be the visible face of networking in the future.

9

The Internet

LO

The first ARPANET message, sent on October 29, 1969, from UCLA to
Stanford. It was supposed to say *LOGIN* but the system crashed.

We've talked about local network technologies like Ethernet and wireless. The
phone system connects telephones worldwide. How do we do the same for computers? How do we scale up to connect one local network to another, perhaps to link all
the Ethernets in a building, or to connect computers in my home to computers in your
building in the next town, or to connect a corporate network in Canada to one in
Europe? How do we make it work when the underlying networks use unrelated technologies? How do we do it in a way that expands gracefully as more networks and
more users are connected, as distances grow, and as equipment and technology
change over time?

The Internet is one answer to those questions, and it has been so successful that
for most purposes, it has become *the* answer.

The Internet is neither a giant network nor a giant computer. It's a loose, unstruc-
tured, chaotic, ad hoc collection of networks, bound together by standards that define
how networks and the computers on them communicate with each other.

How do we connect networks that have different physical properties—optical
fiber, Ethernet, wireless—and that may be far away from each other? We need names
and addresses so we can identify networks and computers, the equivalent of telephone numbers and a phone book. We need to be able to find routes between networks that are not directly connected. We need to agree on how information is formatted as it moves around, and on a great number of other less obvious matters like
coping with errors, delays and overload. Without such agreements, communication
would be difficult or even impossible.

Agreements on data formats, who speaks first and what responses can follow, how
errors are handled, and the like, are dealt with in all networks, and in the Internet
especially, with *protocols*. "Protocol" has somewhat the same meaning as it does in
ordinary speech—a set of rules for interacting with another party—but network

protocols are based on technical considerations, not social custom, and are much more precise than even the most rigid social structure.

It might not be entirely obvious, but the Internet strongly requires such rules: everyone has to agree on protocols and standards for how information is formatted, how it is exchanged between computers, how computers are identified and authorized, and what to do when something fails. Agreeing on protocols and standards can be complicated, since there are many vested interests, including companies that make equipment or sell services, entities that hold patents or secrets, and governments that might want to monitor and control what passes across their borders and between their citizens.

Some resources are in scarce supply; spectrum for wireless services is one obvious example. Names for web sites can't be handled by anarchy. Who allocates such resources and on what basis? Who pays what to whom for the use of limited resources? Who adjudicates the inevitable disputes? What legal system(s) will be used to resolve disputes? Indeed, who gets to make the rules? It might be governments, companies, industry consortia, and nominally disinterested or neutral bodies like the UN's International Telecommunication Union, but in the end everyone has to agree to abide by the rules.

It's clear that such issues can be resolved—after all, the telephone system works worldwide, connecting disparate equipment in different countries. The Internet is more of the same, though it's newer, larger, much more disorderly, and changing more rapidly. It's a free-for-all compared to the controlled environments of traditional phone companies, most of which were either government monopolies or tightly regulated companies. But under government and commercial pressures, the Internet is less free-wheeling and more constrained than in its early days.

9.1 An Internet Overview

Before we dive into details, here's the big picture. The Internet began in the 1960s with attempts to build a network that could connect computers at widely separated geographical locations. The funding for much of this work came from the Advanced Research Projects Agency of the US Department of Defense, and the resulting network came to be called ARPANET. The first ARPANET message was sent from a computer at UCLA to one at Stanford, a distance of about 350 miles or 550 km, on October 29, 1969, which thus could be called the birthday of the Internet. (The bug that caused the initial failure was quickly fixed and the next attempt worked.)

From the beginning, the ARPANET was designed to be robust in the face of failure of any of its components and to route traffic around problems. The original ARPANET computers and technology were replaced over time. The network itself originally linked university computer science departments and research institutions, then spread into the commercial world in the 1990s, becoming "the Internet" somewhere along the way.

Today, the Internet consists of many millions of loosely connected independent networks. Nearby computers are connected by local area networks, often wireless

Ethernets. Networks in turn are connected to other networks via *gateways* or *routers*—specialized computers that route packets of information from one network to the next. (Wikipedia says that a gateway is a more general device and a router is a special case, but the usage is not universal.) Gateways exchange routing information so they know, at least in a local way, what is connected and thus reachable.

Each network may connect many host systems, like computers and phones in homes, offices and dorm rooms. Individual computers within a home are likely to use wireless to connect to a router that in turn is connected to an *Internet Service Provider* or *ISP* by cable or DSL; office computers might use wired Ethernet connections.

As I mentioned in the previous chapter, information travels through networks in chunks called packets. A packet is a sequence of bytes with a specified format; different devices use different packet formats. Part of the packet will contain address information that tells where the packet comes from and where it is going. Other parts of the packet will contain information about the packet itself, like its length, and finally the information being carried, the *payload*.

On the Internet, data is carried in *IP packets* (for "Internet Protocol"). IP packets all have the same format. On any particular network, an IP packet may be transported in one or more physical packets. For instance, a large IP packet will be split into multiple smaller Ethernet packets because the largest possible Ethernet packet (about 1,500 bytes) is much smaller than the largest IP packet (somewhat more than 65,000 bytes).

Each IP packet passes through multiple gateways; each gateway sends the packet on to a gateway that is closer to the ultimate destination. As a packet travels from here to there, it might pass through 20 gateways, owned and operated by a dozen different companies or institutions, and possibly in different countries. Traffic need not follow the shortest path; convenience and cost could route packets through longer routes. Many packets with origins and destinations outside the United States use cables that pass through the US, a fact that the NSA exploits to record worldwide traffic.

To make this work, we need several mechanisms.

Addresses: Each host computer must have an address that will identify it uniquely among all hosts on the Internet, rather like a telephone number. This identifying number, the *IP address*, is either 32 bits (4 bytes) or 128 bits (16 bytes). The shorter addresses are for version 4 of the Internet protocol ("IPv4"), while the longer ones are for version 6 ("IPv6"). IPv4 has been used for many years and is still dominant, but all available IPv4 addresses have now been allocated, so the shift to IPv6 is accelerating.

IP addresses are analogous to Ethernet addresses. An IPv4 address is conventionally written as the values of its 4 bytes, each as a decimal number, separated by periods, as in 140.180.223.42 (which is `www.princeton.edu`). This odd notation is called *dotted decimal*; it's used because it's easier for humans to remember than pure decimal or hexadecimal would be. Figure 9.1 shows that IP address in dotted decimal, binary, and hex.

IPv6 addresses are conventionally written as 16 hexadecimal bytes with colons separating each pair, like `2620:0:1003:100c:9227:e4ff:fee9:05ec`.

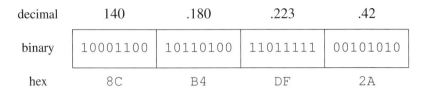

Figure 9.1: Dotted decimal notation for IPv4 addresses.

These are even less intuitive than dotted decimal, so I'll use IPv4 for illustrations. You can determine your own IP address through System Preferences on macOS or an analogous application on Windows, or through Settings on your phone if you're using Wi-Fi.

A central authority assigns a block of consecutive IP addresses to the administrator of a network, which in turn assigns individual addresses to host computers on that network. Each host computer thus has a unique address that is assigned locally according to the network it is on. This address might be permanent for a desktop computer, but for mobile devices it's dynamic and changes at least every time the device reconnects to the Internet.

Names: A host that people will try to access directly must have a name for human use, since few of us are good at remembering arbitrary 32-bit numbers, even in dotted decimal. Names are the ubiquitous forms like `www.nyu.edu` or `ibm.com`, which are called *domain names*. A critical piece of Internet infrastructure, the *Domain Name System* or *DNS*, converts between names and IP addresses.

Routing: There must be a mechanism for finding a path from source to destination for each packet. This is provided by the gateways mentioned above, which continuously exchange routing information among themselves about what is connected to what, and use that to forward each incoming packet onward to a gateway that is closer to its ultimate destination.

Protocols: Finally, there must be rules and procedures that spell out exactly how all of these and other components interoperate, so that information gets copied successfully from one computer to another.

The core protocol, which is called *IP* (Internet Protocol), defines a uniform transport mechanism and a common format for information in transit. IP packets are carried by different kinds of network hardware using their own protocols.

Above IP, a protocol called *TCP* (Transmission Control Protocol) uses IP to provide a reliable mechanism for sending arbitrarily long sequences of bytes from a source to a destination.

Above TCP, higher-level protocols use TCP to provide the services that we think of as "the Internet," like browsing, mail, file sharing, and so on. There are many other protocols as well. For example, changing IP addresses dynamically is handled by a protocol called DHCP (Dynamic Host Configuration Protocol). All these protocols, taken together, define the Internet.

We will talk more about each of these topics in turn.

9.2 Domain Names and Addresses

Who makes the rules? Who controls the allocation of names and numbers?
Who's in charge? For many years, the Internet was managed by informal cooperation among a small group of technical experts. Much of the Internet's core technology was developed by a loose coalition operating as the *Internet Engineering Task
Force* (*IETF*) to produce designs and documents that describe how things should
work. Technical specifications were (and still are) hammered out through regular
meetings and frequent publications called *Requests for Comments*, or *RFCs*, that
eventually become standards. RFCs are available on the web (there are 9,000 of
them). Not all RFCs are deadly serious; check out RFC-1149, "A Standard for the
Transmission of IP Datagrams on Avian Carriers," published on April 1, 1990.

Other aspects of the Internet are managed by an organization called *ICANN*, the
Internet Corporation for Assigned Names and Numbers (`icann.org`), which provides technical coordination of the Internet. That includes assignment of names and
numbers that have to be unique for the Internet to function, like domain names, IP
addresses, and some protocol information. ICANN also accredits domain name
registrars, who in turn assign domain names to individuals and organizations.
ICANN began as an agency of the US Department of Commerce but is now an independent non-profit organization based in California and financed largely by fees from
registrars and domain name registrations.

Not surprisingly, complicated political issues surround ICANN. Some countries
are unhappy with its origin and current location in the US, calling it a tool of the US
government, and there are bureaucrats who would like to see it become part of the
United Nations or another international body, where it could be more easily controlled.

In early 2020, a mysterious private equity group called Ethos Capital made a bid
to take over the `.org` registry, and ICANN agreed to the sale. It became clear that
the goal was to gain control, then raise prices while selling customer data. Fortunately there was a public outcry sufficiently loud that even the Attorney General of
California threatened action, ICANN backed off, and the deal was canceled.

9.2.1 Domain Name System

The Domain Name System or DNS provides the familiar hierarchical naming
scheme that brings us `berkeley.edu` or `cnn.com`. The names `.com`, `.edu` and
so on, and two-letter country codes like `.us` and `.ca`, are called *top-level domains*.
Top-level domains delegate responsibility for administration and further names to
lower levels. For example, Princeton University is responsible for administering
`princeton.edu` and can define sub-domain names within that range, such as
`classics.princeton.edu` for the Classics department and `cs.prince-
ton.edu` for the Computer Science department. Those in turn can define domain
names like `www.cs.princeton.edu` and so on.

Domain names impose a logical structure but need not have any geographical significance. For instance, IBM operates in many countries but its computers are all
included in `ibm.com`. It's possible for a single computer to serve multiple domains,
which is common for companies that provide hosting services; conversely, a single

domain may be served by many computers, as with large sites like Facebook and Amazon.

The lack of geographical constraints has some interesting implications. For example, the country of Tuvalu (population 11,000), a group of small islands in the South Pacific halfway between Hawaii and Australia, has the country code `.tv`. Tuvalu leases the rights to that country code to commercial interests, who will happily sell you a `.tv` domain. If the name you want has some potential commercial significance, say `news.tv`, you'll likely have to pay handsomely for it. On the other hand, `kernighan.tv` is under thirty dollars a year. Other countries blessed by linguistic accidents include the Republic of Moldova, whose `.md` might appeal to doctors, and Italy, which shows up in sites like `play.it`. Normally domain names are restricted to the 26 letters of English, digits and hyphens, but in 2009, ICANN approved some internationalized top-level domain names like

. 中国

as an alternative to `.cn` for China, and

. مصر

in addition to `.eg` for Egypt.

Around 2013, ICANN began to authorize new top-level domains like `.online` and `.club`. It's not clear how successful these will be in the long run, though some, like `.info` and `.io`, seem to be popular. Commercial and governmental domains like `.toyota` and `.paris` are also available for a price, and have raised questions about ICANN's motives—are such domains necessary or are they merely a way to generate more revenue?

9.2.2 IP addresses

Each network and each connected host computer must have an IP address so that it can communicate with others. An IPv4 address is a unique 32-bit quantity: only one host at a time can use that value over the whole Internet. The addresses are assigned in blocks by ICANN, and the blocks are sub-allocated by the institutions that receive them. For example, Princeton has two blocks, 128.112.ddd.ddd and 140.180.ddd.ddd, where each ddd is a decimal number between 0 and 255. Each of these blocks allows up to 65,536 (2^{16}) hosts, or about 131,000 in all.

These blocks of addresses have no numerical or geographical significance whatsoever. Just as the numerically adjacent US telephone area codes 212 and 213 are New York City and Los Angeles, a continent apart, there's no reason to expect that adjacent blocks of IP addresses represent physically nearby computers, and there's no way to infer a geographical location from an IP address by itself, though it's often possible to deduce where an IP address is from other information. For example, DNS supports reverse lookups (IP address to name), and reports that 140.180.223.42 is `www.princeton.edu`, so it's reasonable to guess that's in Princeton, New Jersey, though the server could be somewhere else entirely.

It is sometimes possible to learn more about who is behind a domain name using a service called `whois`, available on the web at `whois.icann.org` or as the Unix command-line program `whois`.

There are only 2^{32} possible IPv4 addresses, which is about 4.3 billion. That's less than one per person on earth, so at the rate that people are using more and more communications services, something will run out. In fact, the situation is worse than it sounds, since IP addresses are handed out in blocks and thus are not used as efficiently as they might be (are there 131,000 computers concurrently active at Princeton?). In any case, with a few exceptions, all IPv4 addresses have been allocated in most parts of the world.

Techniques for piggy-backing multiple hosts onto a single IP address provide some breathing room. Home wireless routers generally use *network address translation* or *NAT*, where a single external IP address can serve multiple internal IP addresses. If you have a NAT, all of your home devices appear externally to have the same IP address; hardware and software in the device handle the conversion in both directions. For example, in my house there are at least a dozen computers and other gadgets that need an IP address. They are all served by a NAT with a single external address.

Once the world shifts to IPv6, which uses 128-bit addresses, the pressure will be off—there are 2^{128} or about 3×10^{38} of those, so we won't run out in a hurry.

9.2.3 Root servers

The crucial DNS service is to convert names into IP addresses. The top-level domains are handled by a set of *root name servers* that know the IP addresses for all top-level domains, for example `mit.edu`. To determine the IP address for `www.cs.mit.edu`, one asks a root server for `mit.edu`'s IP address. That's sufficient to get to MIT, where one asks the MIT name server for `cs.mit.edu`, and that in turn leads to a name server that knows about `www.cs.mit.edu`.

DNS thus uses an efficient algorithm for searching: an initial query at the top immediately eliminates most potential addresses from further consideration. The same is true at each level as the search works its way down the tree. It's the same idea as we saw earlier in hierarchical file systems.

In practice, name servers maintain caches of names and addresses that have been looked up and passed through them recently, so a new request can often be answered with local information rather than going far away. If I access `kernighan.com`, the odds are strong that no one else has looked at that recently, and the local name servers may have to ask a root server for the IP address. If I use that name again soon, however, the IP address is cached nearby and the query runs faster. When I tried this, the first query took a quarter of a second; a few seconds later the same query took less than one tenth of that, as did another query several minutes later.

You can do your own DNS experiments with commands like `nslookup`. Try the Unix command

```
nslookup a.root-servers.net
```

In principle one could imagine having a single root server, but it would be a single point of failure, truly a bad idea for such a critical system. So there are thirteen root servers, spread over the whole world, with about half in the US. Most of these servers consist of multiple computers at widely separated locations that function like a single computer but use a protocol that routes requests to a nearby member of the group. Root servers run different software systems on different kinds of hardware, so they are less vulnerable to bugs and viruses than a monoculture would be. Nevertheless, from time to time root servers are subjected to coordinated attacks and it's conceivable that some combination of circumstances could bring them all down at once.

9.2.4 Registering your own domain

It's easy to register your own domain if the name you want isn't already taken. ICANN has accredited hundreds of registrars around the world, so you can pick one, choose your domain name, pay for it, and it's yours (though you have to renew every year). There are some restrictions but there seem to be no rules against obscenities (easily verified by trying a few) or personal attacks, to the point where corporations and public figures are forced to preemptively acquire domains like `bigcorp-sucks.com` in self-defense. Names are limited to 63 characters and normally only contain letters, digits and hyphens, though it is possible to use Unicode characters; if there are non-ASCII characters, a standard encoding called Punycode converts back into the letter-digit-hyphen subset.

You need a *host* for your site, that is, a computer to hold and serve the content that your site will display to visitors. You also need a *name server* to respond with the IP address of your host when someone tries to find your domain's IP address. That's a separate component, though registrars usually provide the service or easy access to someone who does.

Competition keeps prices down; a `.com` registration costs ten or twenty dollars initially and a similar amount per year to maintain. A hosting service costs five or ten dollars a month for low-volume casual use; merely "parking" a domain with a generic page might well be free. Some hosting services are free, or have a nominal price if you're not doing much with it, or for a short time while you're testing the waters.

Who owns a domain name? How are disputes resolved? What can I do if someone else has registered `kernighan.com`? That last one is easy: not much except offer to buy it. For a name with commercial value, like `mcdonalds.com` or `apple.com`, courts and ICANN's dispute resolution policies have tended to favor the party with the clout. If your name is McDonald or Apple, you won't have much chance of wresting a domain away from one of them, and you might have trouble hanging on to it even if you got there first. (In 2003, a Canadian high school student named Mike Rowe set up a web site at `mikerowesoft.com` for his tiny software business. This brought a threat of legal action from a rather larger corporation with a similar-sounding name. Eventually the case was settled and Mr. Rowe chose a different domain name.)

9.3 Routing

Routing—finding a path from source to destination—is a central problem in any large network. Some networks use static routing tables that provide a next step in the path for all possible destinations. The problem with the Internet is that it's far too large and dynamic for static tables. As a result, Internet gateways continuously refresh their routing information by exchanging information with adjacent gateways; this ensures that information about possible and desirable paths is always relatively up to date.

The sheer scale of the Internet requires a hierarchical organization to manage routing information. At the top level, some tens of thousands of *autonomous systems* provide routing information about the networks they contain; normally an autonomous system corresponds to a large Internet service provider (ISP). Within a single autonomous system, routing information is exchanged locally, but the system presents unified routing information to outside systems.

Although it's not formal or rigorous, there is a sort of physical hierarchy as well. One accesses the Internet via an ISP, which is a company or other organization that in turn connects to other Internet providers. Some ISPs are tiny, some are huge (those run by telephone and cable companies, for example); some are run by organizations like a company, university or government agency, while others offer access as a service for a fee—telephone and cable companies are typical examples. Individuals connect to their ISP by cable (common for residential service) or phone; companies and schools offer Ethernet or wireless connections.

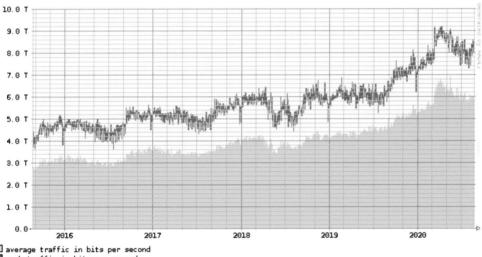

average traffic in bits per second
peak traffic in bits per second
Current 5902.4 G
Averaged 3945.5 G
Graph Peak 9168.5 G
DE-CIX All-Time Peak 9168.55
Created at 2020-08-22 15:23 UTC
Copyright 2020 DE-CIX Management GmbH

Figure 9.2: Traffic at DE-CIX Frankfurt IXP (courtesy of DE-CIX).

ISPs connect to each other through gateways. For high volume between major carriers, there are *Internet exchange points* (IXP) where network connections from multiple companies meet and physical connections are made between networks, so data from one network can be efficiently passed to another. Large exchanges pass terabits per second from one network to another; for example, the DE-CIX Frankfurt exchange, one of the largest in the world, presently averages nearly 6 Tbps and has had peaks above 9 Tbps. Figure 9.2 shows a 5-year graph of traffic. Notice the steady growth, and the significant rise in traffic when the Covid-19 crisis that began early in 2020 forced many people to work remotely.

Some countries have comparatively few gateways that provide access into or out of the country, and these can be used to monitor and filter traffic that the government considers undesirable.

You can explore routing with a program called `traceroute` on Unix systems (including Macs) or `tracert` on Windows, and there are web-based versions as well. Figure 9.3 shows the path from Princeton to a computer at the University of Sydney in Australia, edited for space. Each line shows the name, IP address and round trip time for the next hop in the path.

```
$ traceroute sydney.edu.au
traceroute to sydney.edu.au (129.78.5.8),
          30 hops max, 60 byte packets
 1 switch-core.CS.Princeton.EDU (128.112.155.129) 1.440 ms
 2 csgate.CS.Princeton.EDU (128.112.139.193) 0.617 ms
 3 core-87-router.Princeton.EDU (128.112.12.57) 1.036 ms
 4 border-87-router.Princeton.EDU (128.112.12.142) 0.744 ms
 5 local1.princeton.magpi.net (216.27.98.113) 14.686 ms
 6 216.27.100.18 (216.27.100.18) 11.978 ms
 7 et-5-0-0.104.rtr.atla.net.internet2.edu (198.71.45.6) 20.089 ms
 8 et-10-2-0.105.rtr.hous.net.internet2.edu (198.71.45.13) 48.127 ms
 9 et-5-0-0.111.rtr.losa.net.internet2.edu (198.71.45.21) 75.911 ms
10 aarnet-2-is-jmb.sttlwa.pacificwave.net (207.231.241.4) 107.117 ms
11 et-0-0-1.pe1.a.hnl.aarnet.net.au (202.158.194.109) 158.553 ms
12 et-2-0-0.pe2.brwy.nsw.aarnet.net.au (113.197.15.98) 246.545 ms
13 et-7-3-0.pe1.brwy.nsw.aarnet.net.au (113.197.15.18) 234.717 ms
14 138.44.5.47 (138.44.5.47) 237.130 ms
15 * * *
16 * * *
17 shared-addr.ucc.usyd.edu.au (129.78.5.8) 235.266 ms
```

Figure 9.3: Traceroute from Princeton, NJ, to University of Sydney, Australia.

The round-trip times show a meandering trip across the United States, then two big hops across the Pacific to Australia. It's fun to try to figure out where the various gateways are from the cryptic abbreviations in their names. A connection from one country to another can easily go through gateways in other countries as well, often including the US; this might be surprising, and perhaps undesirable, depending on the nature of the traffic and the countries involved. The map of submarine cables in Figure 9.4 shows the degree to which fiber optic cables reach land in the US, Europe and Asia. The figure does not show cables on land.

Unfortunately, security concerns have made `traceroute` less informative over time, as more and more sites choose not to provide the information necessary to make it work. For instance, some sites don't reveal names or IP addresses; those are marked with asterisks in the figure.

9.4 TCP/IP Protocols

A protocol defines the rules that govern how two parties interact with each other: whether one offers to shake hands, how deeply to bow, who goes through the door first, which side of the road to drive on, and so on. Most protocols in daily life are

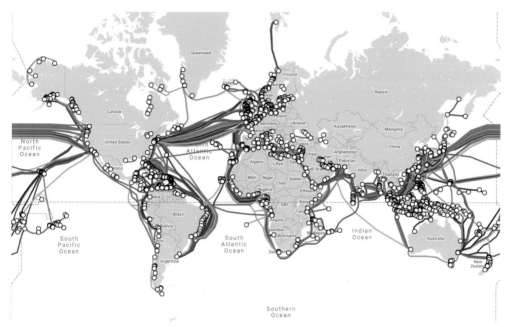

Figure 9.4: Submarine cables (courtesy of submarinecablemap.com).

pretty informal, though the proper side of the road has the force of law. By contrast, network protocols are very precisely specified.

The Internet has many protocols, two of which are absolutely fundamental. IP is the *Internet Protocol*; it defines how individual packets are formatted and transmitted. TCP, the *Transmission Control Protocol*, defines how IP packets can be combined into streams of data and connected to services. Together the pair of protocols is called *TCP/IP*.

Gateways route IP packets, though each physical network has its own format for carrying IP packets. Each gateway has to convert between the network format and IP as packets come in and go out.

Above the IP level, TCP provides reliable communication, so that users (programmers, really) don't have to think about packets, just streams of information. Most of the services that we think of as "the Internet" use TCP.

Above these are application-level protocols that provide those services, mostly built on TCP: the web, mail, file transfer, and the like. Thus there are several layers of protocols, each relying on the services of the one below and providing services to the one above. This is an excellent example of the layering of software described in Chapter 6, and one conventional diagram (Figure 9.5) looks vaguely like a layered wedding cake.

UDP, the *User Datagram Protocol*, is another protocol at the same level as TCP. UDP is much simpler than TCP and is used for data exchange that doesn't require a two-way stream, just efficient packet delivery with a few extra features. DNS uses UDP, as do video streaming, voice over IP, and some online games.

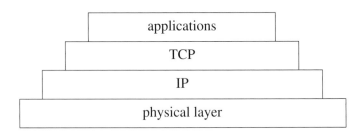

Figure 9.5: Protocol layers.

9.4.1 IP, the Internet Protocol

IP, the Internet Protocol, provides an unreliable, connectionless packet delivery service. "Connectionless" means that each IP packet is self-contained and has no relationship to any other IP packet. IP has no state or memory: the protocol does not have to remember anything about a packet once it has been passed on to the next gateway.

"Unreliable" means both more and less than it appears to. IP is a "best effort" protocol that makes no guarantees about how well it delivers packets—if something goes wrong, that's tough. Packets may be lost or damaged, they may be delivered out of order, and they may arrive too quickly to be processed or too slowly to be useful. In actual use, IP is very reliable. When a packet does go astray or get damaged, however, there's no attempt at recovery. It's like dropping a postcard into a mailbox in some strange place: the postcard probably gets delivered though it might be damaged en route. Sometimes it doesn't arrive at all, and sometimes delivery takes much longer than you expected. (There is one IP failure mode that postcards don't share: an IP packet can be duplicated, so the recipient gets more than one copy.)

IP packets have a maximum size of about 65 KB. Thus a long message must be broken up into smaller chunks that are sent separately, then reassembled at the far end. An IP packet, like an Ethernet packet, has a specified format. Figure 9.6 shows some of the IPv4 format; the IPv6 packet format is analogous, but the source and destination addresses are each 128 bits long.

version	type	hdr len	total len	TTL	source address	destination address	error check	data (up to 65KB)

Figure 9.6: IPv4 packet format.

One interesting part of the packet is the *time to live*, or TTL. The TTL is a one-byte field that is set to an initial value (typically about 40) by the source of the packet and is decreased by one by each gateway that handles the packet. If the count ever gets down to zero, the packet is discarded and an error packet is sent back to the originator. A typical trip through the Internet might involve 15 to 20 gateways, so a packet that takes 255 hops is clearly in trouble, probably in a loop. The TTL field doesn't eliminate loops but it does prevent individual packets from living forever.

The IP protocol itself makes no guarantees about how fast data will flow: as a best-effort service, it doesn't even promise that information will arrive, let alone how fast. The Internet makes extensive use of caching to try to keep things moving; we've already seen this in the discussion of name servers. Web browsers also cache information, so if you try to access a page or an image that you've looked at recently, it may come from a local cache rather than from the network. Major Internet servers also use caching to speed up responses. Companies like Akamai provide content distribution services for other companies like Yahoo; this amounts to caching content closer to the recipients. Search engines also maintain large caches of pages they have found during their crawl of the web; that's a topic for Chapter 11.

9.4.2 TCP, the Transmission Control Protocol

Higher-level protocols synthesize reliable communications from this unreliable substrate. The most important of these is TCP, the Transmission Control Protocol. To its users, TCP provides a reliable two-way stream: put data in at one end and it comes out the other end, with little delay and low probability of error, as if it were a direct wire from one end to the other.

I won't go into the details of how TCP works—there are a lot of them—but the basic idea is simple enough. A stream of bytes is chopped up into pieces and put into TCP packets or *segments*. A TCP segment contains not only the real data but also a "header" with control information, including a sequence number so the recipient knows which part of the stream each packet represents; in this way a lost segment will be noticed and can be resent. Error-detection information is included, so if a segment is damaged, that is likely to be detected as well. Each TCP segment is transmitted in an IP packet. Figure 9.7 shows the contents of a TCP segment header, which will be sent inside an IP packet along with the data.

source port	destination port	sequence number	acknowledgment	error check	other info

Figure 9.7: TCP segment header format.

Each segment must be acknowledged, positively or negatively, by the recipient. For each segment that I send to you, you must send me an acknowledgment that you got it. If I don't get the acknowledgment after a decent interval, I must assume that the segment was lost and I will send it again. Similarly, if you are expecting a particular segment and haven't received it, you must send me a *negative acknowledgment* ("Segment 27 has not arrived") and I will know to send it again.

Of course, if the acknowledgments themselves are lost, the situation is even more complicated. TCP has a number of timers that determine how long to wait before assuming that something is wrong. If an operation is taking too long, recovery can be attempted; ultimately, a connection will time out and be abandoned. (You've probably encountered this with unresponsive web sites.) All of this is part of the protocol.

The protocol also has mechanisms to make this work efficiently. For example, a sender can send packets without waiting for acknowledgments of previous packets, and a receiver can send a single acknowledgment for a group of packets; if traffic is flowing smoothly, this reduces the overhead due to acknowledgments. If congestion occurs and packets begin to get lost, however, the sender quickly backs off to a lower rate, and only slowly builds back up again.

When a TCP connection is set up between two host computers, the connection is not just to a specific computer but to a specific *port* on that computer. Each port represents a separate conversation. A port is represented by a two-byte (16-bit) number, so there are 65,536 possible ports, and thus in principle a host could be carrying on 65,536 separate TCP conversations at the same time. This is analogous to a company having a single phone number and employees having different extensions.

A hundred or so "well known" ports are reserved for connections to standard services. For instance, web servers use port 80, and mail servers use port 25. If a browser wants to access `www.yahoo.com`, it will establish a TCP connection to port 80 at Yahoo, but a mail program will use port 25 to access a Yahoo mail server. The source and destination ports are part of the TCP header that accompanies the data.

There are many more details but the basic ideas are no more complicated than this. TCP and IP were originally designed around 1973 by Vinton Cerf and Robert Kahn, who shared the 2004 Turing Award for their work. Although the TCP/IP protocols have undergone refinements, they remain essentially the same even though network sizes and traffic speeds have grown by many orders of magnitude. The original design was remarkably well done, and today TCP/IP handles most of the traffic on the Internet.

9.5 Higher-Level Protocols

TCP provides a reliable two-way stream that transfers data back and forth between two computers. Internet services and applications use TCP as the transport mechanism but have their own protocols specific to the task at hand. For example, HTTP, the Hypertext Transfer Protocol, is a particularly simple protocol used by web browsers and servers. When I click on a link, my browser opens a TCP/IP connection to port 80 on the server, say `amazon.com`, and sends a short message that requests a specific page. In Figure 9.8, the browser is the client application at the top left; the message goes down the chain of protocols, crosses the Internet (normally with many more steps), and then back up to the matching server application at the far end.

At Amazon, the server prepares the page, then sends it to me along with a small amount of additional data, perhaps information about how the page is encoded; the return path need not be the same as the original path. My browser reads this response and uses the information to display the contents.

9.5.1 Telnet and SSH: remote login

Given the Internet as a carrier of information, what can we do with it? We'll look at a couple of the earliest TCP/IP applications that used the nascent Internet; they date from the early 1970s, yet are still used today, a tribute to their design and utility.

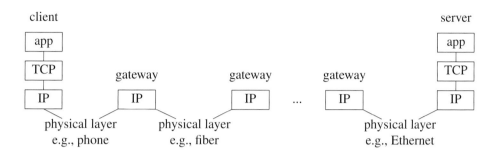

Figure 9.8: TCP/IP connections and information flow.

They are command-line programs and although mostly simple to use, they are aimed at comparative experts, not casual users.

You can access Amazon by using Telnet, a TCP service for establishing a remote login session on another machine. Normally Telnet uses port 23, but it can be aimed at other ports as well. Type these lines into a command-line window:

```
$ telnet www.amazon.com 80
GET / HTTP/1.0
    [type an extra blank line here]
```

and back will come over 225,000 characters that a browser would use to display the page.

GET is one of a handful of HTTP requests, "/" asks for the default file at the server, and HTTP/1.0 is the protocol name and version. We'll talk more about HTTP and the web in the next chapter.

Telnet provides a way to access a remote computer as if one were directly connected to it. Telnet accepts keystrokes from the client and passes them to the server as if they had been typed there directly; it intercepts the server's output and sends it back to the client. Telnet makes it possible to use any computer on the Internet as if it were local, if one has the right permissions. As another example, here's how to use it to do a search:

```
$ telnet www.google.com 80
GET /search?q=whatever
    [type an extra blank line here]
```

This produces over 110,000 bytes of output, mostly JavaScript and images, but the search results are visible within if one looks carefully.

Telnet offers no security. If the remote system accepts logins without a password, none is requested. If the remote system asks for a password, Telnet sends the client's password in the clear, so anyone watching the flow of data could see it. This total lack of security is one reason why Telnet is now rarely used except in special circumstances where security does not matter. Its descendant *SSH* (Secure Shell) is widely used, however, since it encrypts all traffic in both directions and thus can exchange information securely; it uses port 22.

9.5.2 SMTP: Simple Mail Transfer Protocol

The second protocol is *SMTP*, the *Simple Mail Transfer Protocol*. We usually send and receive mail with a browser or a standalone program. But like many other things on the Internet, several layers lie beneath this surface, each enabled by programs and protocols. Mail involves two basic kinds of protocols. SMTP is used to exchange mail with another system. It establishes a TCP/IP connection to port 25 at the recipient's mail computer and uses the protocol to identify the sender and recipient, and to transfer messages. SMTP is text based; you can run it with Telnet on port 25 if you want to see how it works, though there are enough security restrictions that you might have trouble using it even locally on your own computer. Figure 9.9 shows a sample dialog (edited for compactness) from an actual session with a local system, in which I sent myself mail as if from someone else (in effect, spam). My typing is in **bold italic**.

```
$ telnet localhost 25
Connected to localhost.
220 davisson.princeton.edu ESMTP Postfix
HELO localhost
250 davisson.princeton.edu
mail from:liz@royal.gov.uk
250 2.1.0 Ok
rcpt to:bwk@princeton.edu
250 2.1.0 Ok
data
354 End data with <CR><LF>.<CR><LF>
Subject: knighthood?

Dear Brian --

Would you like to be knighted?  Please let me know soon.

ER
.
250 2.0.0 p4PCJfD4030324 Message accepted for delivery
quit
```

Figure 9.9: Sending mail with SMTP.

This nonsensical (or at least improbable) message was duly delivered to my mailbox, as seen in Figure 9.10.

Since SMTP requires that mail messages be ASCII text, a standard called MIME (Multipurpose Internet Mail Extensions, in effect another protocol) describes how to convert other kinds of data into text and how to combine multiple pieces into a single mail message. This is the mechanism used to include mail attachments like pictures and video, and it's also used by HTTP.

Although SMTP is an end-to-end protocol, the TCP/IP packets pass through 15 to 20 gateways on their way from source to destination. It is entirely possible for any gateway along the path to examine the packets and make copies for leisurely inspection. SMTP itself can make copies of the contents, and mail systems keep track of contents and headers. If the contents are to be kept private, they must be encrypted at

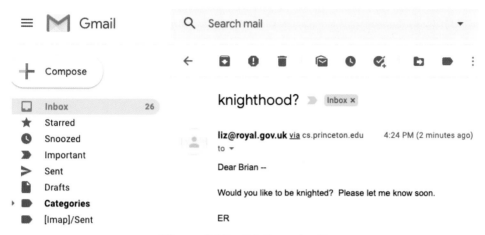

Figure 9.10: Mail received!

the source. Bear in mind that encrypting the contents does not conceal the identities of sender and receiver. Traffic analysis reveals who is communicating with whom; such metadata is often as informative as the actual contents, as we will discuss in Chapter 11.

SMTP transfers mail from source to destination but it has nothing to do with access to the mail thereafter. Once the mail arrives at the target computer, it normally waits there until the recipient retrieves it, usually with another protocol called *IMAP* (Internet Message Access Protocol). With IMAP, your mail remains on a server and you can access it from multiple places. IMAP ensures that your mailbox is always in a consistent state even if there are multiple simultaneous readers and updaters, as is the case when you handle your mail from a browser and a phone. There's no need to make multiple copies of messages or to copy them around among computers.

It's common for mail to be handled "in the cloud" by systems like Gmail or Outlook. Underneath, these use SMTP for transfer and behave like IMAP for client access. I'll talk about cloud computing in Chapter 11.

9.5.3 File sharing and peer-to-peer protocols

In June 1999, Shawn Fanning, a freshman at Northeastern University, released Napster, a program that made it dead easy for people to share music compressed in MP3 format. Fanning's timing was great. Audio CDs of popular music were ubiquitous but expensive. Personal computers were fast enough to do MP3 encoding and decoding, and the algorithms were widely available. Bandwidths were high enough that songs in MP3 format could be transferred across a network reasonably quickly, especially for college students on dormitory Ethernets. Fanning's design and implementation were well done and Napster spread like wildfire. A company was formed to provide the service in mid-1999, and claimed to have 80 million users at its peak. The first lawsuits were filed later in 1999, alleging theft of copyrighted music on a grand scale, and a court decision put Napster out of business by mid-2001. From nothing to 80 million users to nothing in barely two years was a vivid illustration of the then-popular phrase "Internet time."

To use Napster, one had to download a Napster client program to run on one's own computer. The client established a local folder for the files that would be shareable. When the client subsequently logged in to a Napster server, it uploaded the *names* of shareable files and Napster added them to a central directory of currently available filenames. The central directory was continuously updated: as new clients connected, their filenames were added, and when a client failed to respond to a probe, its filenames were de-listed.

When a user searched the central directory for song titles or performers, Napster supplied a list of other users who were currently online and willing to share those files. When a user selected a supplier, Napster arranged the contact (sort of like a dating service) by providing an IP address and a port number, and the client program on the user's computer made direct contact with the supplier and retrieved the file. The supplier and the consumer reported status to Napster, but the central server was otherwise "not involved" because it never touched any of the music itself.

We are used to the client-server model, where a browser (the client) requests something from a web site (the server). Napster was an example of a different model. It provided a central directory that listed the music that was currently available for sharing, but the music itself was only stored on user machines, and when a file was transferred, it went directly from one Napster user to another, rather than through the central system. Hence the organization was called *peer-to-peer*, with the sharers being the peers. Because the music itself was only stored on the peer computers, never on the central server, Napster hoped to sidestep copyright issues, but courts were not persuaded by that legal fine point.

The Napster protocol used TCP/IP, so it was in effect a protocol at the same level as HTTP and SMTP. Without in any way deprecating Fanning's work, which was neat indeed, Napster is a simple system when the infrastructure of the Internet, TCP/IP, MP3, and tools for building graphical user interfaces are in place already.

Most current file sharing, whether legal or not, uses a peer-to-peer protocol called BitTorrent, which was developed by Bram Cohen in 2001. BitTorrent is particularly good for sharing large popular files like movies and TV programs, because each site that starts to download a file with BitTorrent must also begin to upload pieces of the file to others who want to download. Files are found by searching distributed directories and a small "torrent file" is used to identify a tracker that maintains a record of who has sent and received which blocks. BitTorrent users are vulnerable to detection, since the protocol requires that downloaders also upload and thus they are easily identified in the act of making allegedly copyrighted material available.

Peer-to-peer networks have other uses than file sharing of questionable legality. Bitcoin, a digital currency and payment system that we will discuss in Chapter 13, also uses a peer-to-peer protocol.

9.6 Copyright on the Internet

In the 1950s, it was not practical to copy a book or an audio recording. Copying steadily became cheaper, however, and by the 1990s it was easy to make a digital copy of a book or a record, and those copies could be made in quantity and sent to others via the Internet at high speed and zero cost.

   The entertainment industry, through trade groups like the Recording Industry Association of America (RIAA) and the Motion Picture Association of America (MPAA), has been tireless in trying to prevent sharing of copyrighted material. This has included lawsuits and threats of legal action against large numbers of alleged copyright infringers, as well as intense lobbying efforts in support of legislation to make such activities illegal. Piracy will likely always be with us, but it appears that by charging reasonable prices for guaranteed quality, businesses can greatly reduce its effects and still make money; Apple's iTunes music store is an example, as are streaming services like Netflix and Spotify.

   In the US, the primary law for digital copyright issues is the 1998 Digital Millennium Copyright Act or DMCA, which made it illegal to circumvent copyright protection techniques on digital media, including distribution of copyrighted material on the Internet; other countries have analogous laws. The DMCA is the legal mechanism used by the entertainment industry to go after copyright violators.

   The DMCA provides a "safe harbor" provision for Internet service providers: if an ISP is notified by a legitimate copyright holder that a user of the ISP is supplying copyrighted material, the ISP itself is not liable for copyright violation if it requires the infringer to remove the copyrighted material. This safe harbor provision is important at universities, where the university is the ISP for its students and faculty. Thus every university has some official who deals with allegations of infringement. Figure 9.11 is the DMCA notice for Princeton.

   To report copyright infringements involving Princeton University information technology resources or services, please notify [...], the agent designated under the Digital Millennium Copyright Act, P.L. 105-304, to respond to reports alleging copyright infringements on Princeton University website locations.

Figure 9.11: DMCA notification information from a web page.

   The DMCA is also invoked (on both sides) in legal tussles between more evenly matched opponents. In 2007 Viacom, a major film and TV company, sued Google for $1 billion over copyrighted material available on YouTube, a Google service. Viacom said that the DMCA was not meant to enable wholesale theft of copyrighted material. Part of Google's defense was that it responded appropriately to DMCA take-down notices when they were properly presented, but that Viacom had not done so. A judge ruled in favor of Google in June 2010, an appeals court reversed part of the decision, then another judge ruled in favor of Google, again on the grounds that YouTube was following DMCA procedures properly. The parties settled in 2014 but regrettably the terms of the settlement were not made public.

   In 2004, Google began a project to scan a large number of books held primarily in research libraries. In 2005 it was sued by the Authors Guild, which alleged that Google was profiting by violating the copyrights of authors. The case dragged on for a very long time, but a decision in 2013 said that Google was not guilty, on the grounds that it preserved books that might otherwise be lost, made them available in digital form for scholarship, and could even generate income for authors and publishers. An appeals court affirmed this decision late in 2015, based in part on the fact that Google made only limited amounts of each book available online. The Authors Guild appealed to the Supreme Court, which in 2016 declined to hear the case, thus effectively ending the dispute.

This is another case where one can see reasonable arguments on both sides. As a researcher I would like to be able to search in books that I might otherwise not be able to see or even know about, but as an author I want people to buy legitimate copies of my books rather than download pirated copies.

It's easy to file DMCA complaints. I sent one to Scribd about an illegal uploaded copy of the first edition of this book, and they took it down within 24 hours. Unfortunately, most illegal copies of most books are basically impossible to remove.

The DMCA is sometimes used in an anti-competitive way that was presumably not part of its original intent. For example, Philips makes "smart" network-connected light bulbs that allow controllers to adjust their brightness and color. Late in 2015, Philips announced that it was modifying the firmware so that only Philips bulbs could be used with Philips controllers. The DMCA would prevent anyone else from reverse-engineering the software to allow third-party bulbs. There was a considerable outcry and Philips backed down for this specific case, but other companies continue to use the DMCA to limit competition, for example with replacement cartridges for printers and coffee makers.

9.7 The Internet of Things

Smartphones are just computers that are able to use the standard phone system, but all modern phones can also access the Internet either via the wireless carrier or through Wi-Fi if it's available. This accessibility blurs the distinction between the telephone network and the Internet, a distinction that's likely to eventually fade away.

The same forces that have made mobile phones such a pervasive part of our world today operate on other digital devices as well. As I've said earlier, many gadgets and devices contain powerful processors, memory, and often a wireless network connection as well. It's natural to want to connect such devices to the Internet, and it's easy because all the necessary mechanisms are already in place, and the incremental cost is close to zero. Thus we see cameras that can upload pictures by Wi-Fi or Bluetooth, cars that download entertainment while uploading location and engine telemetry, thermostats that measure and control their environment and report to the absent home-owner, video monitors that keep an eye on children and nannies and people who ring the doorbell, voice response systems like Alexa, and as mentioned above, networked light bulbs, all based on Internet connections. The popular buzzword for all of these is the *Internet of Things*, or *IoT*.

In many respects, this is a great idea and it's certain that the future holds more and more of it. But there is a large downside as well. These specialized devices are more vulnerable to problems than their general-purpose peers. Hacking, break-ins, damage, and so on are quite possible, and in fact are more likely because attention to security and privacy for the Internet of Things has lagged well behind the state of the art for personal computers and phones. A surprising number of devices "call home" by sending information back to servers in their country of manufacture.

As one random example from a rich selection, in January 2016 a web site allowed its users to search for web cameras that display their video without any protection at all. The site offers images for "marijuana plantations, back rooms of banks, children, kitchens, living rooms, garages, front gardens, back gardens, ski slopes, swimming pools, colleges and schools, laboratories, and cash register cameras in retail stores." One can imagine uses from simple voyeurism to much worse.

Some children's toys are Internet-enabled, which opens up a different set of potential hazards. One study showed that several toys included analytics code that could be used to track children, and insecure mechanisms that would permit using the toy as a vector for other attacks. (One of the toys was an Internet-connected water bottle apparently intended to monitor hydration.) The potential tracking is a violation of COPPA, the Children's Online Privacy Protection Act, and of the stated privacy policies of the toys.

Consumer products like the web cams mentioned above are often vulnerable because manufacturers don't provide good security. It may be deemed too costly to bother with, or too complicated for consumers, or it could just be badly implemented. For example, in late 2019, a hacker posted IP addresses and Telnet passwords for half a million IoT devices, which he had found by scanning for devices that responded on port 22, then trying default accounts and passwords like "admin" and "guest."

Infrastructure systems for power, communications, transportation and many other things have been connected to the Internet without enough attention to protecting them. As one instance, in December 2015 it was reported that wind turbines from a particular manufacturer have a web-enabled administrative interface and can be trivially attacked (just by editing the URL) to shut off the power that they are generating.

9.8 Summary

There are only a handful of basic ideas behind the Internet; it is remarkable how much can be accomplished with so little mechanism (though with a great deal of engineering).

The Internet is a packet network: information is sent in standardized individual packets that are routed dynamically through a large and changing collection of networks. This is a different model from the telephone system's circuit network, where each conversation has a dedicated circuit, conceptually a private wire between the two talking parties.

The Internet assigns a unique IP address to each host that is currently connected, and hosts on the same network share a common IP address prefix. Mobile hosts like laptops and phones will likely have a different IP address each time they are connected, and the IP address might change as the host moves around. The Domain Name System is a large distributed database that converts names to IP addresses and vice versa.

Networks are connected by gateways, specialized computers that route packets from one network to the next as the packets make their way to their destination. Gateways exchange routing information, using routing protocols, so they always know how to forward a packet to get it closer to where it is bound, even as network topology changes and connections come and go.

The Internet lives and breathes by protocols and standards. IP is the common mechanism, the lingua franca for exchange of information. Specific hardware technologies like Ethernet and wireless systems encapsulate IP packets, but the details of how any particular piece of hardware works, or even that it's involved, are not visible at the IP level. TCP uses IP to create a reliable stream connected to a specific port at a host. Higher-level protocols use TCP/IP to create services.

Protocols divide the system into layers. Each layer uses the services provided by the layer immediately below it and provides services to the layer directly above it; no layer tries to do everything. This layering of protocols is fundamental to the operation of the Internet; it's a way to organize and control complexity while hiding irrelevant details of implementation. Each layer sticks to what it knows how to do—hardware networks move bytes from one computer on a network to another, IP moves individual packets across the Internet, TCP synthesizes a reliable stream from IP, and application protocols send data back and forth on the stream. The programming interfaces that each layer presents are fine examples of the APIs we talked about in Chapter 5.

What these protocols have in common is that they move information between computer programs, using the Internet as a dumb network that copies bytes efficiently from one computer to another without trying to interpret or process them. This is an important property of the Internet: it's "dumb" in the sense that it leaves the data alone. Less pejoratively, this is known as the *end-to-end principle*: the intelligence lies at the end points, that is, with the programs that send and receive the data. This contrasts with traditional telephone networks, where all the intelligence has been in the network, and the end points, like old-fashioned telephones, were truly dumb, providing little more than a way to connect to the network and relay voice.

The "dumb network" model has been highly productive, since it means that anyone with a good idea can create smart end points and rely on the network to carry the bytes; waiting for a telephone or cable company to implement or support the good idea would not work. As might be expected, carriers would be happier with more control, especially in the mobile area, where most of the innovation comes from elsewhere. Smartphones like iPhone and Android are computers that primarily communicate over the telephone network instead of the Internet. Carriers would love to make money from telephone services, but basically can only derive revenue from carrying data. In the early days, most cell phones had a flat rate per month for data service, but at least in the US that long ago changed to a structure that charges more for more usage. For high-volume services like downloading movies, higher prices and caps for truly abusive use might be reasonable, but it seems less defensible for services like texting, which costs the carriers almost nothing since the bandwidth is so tiny.

Finally, notice how the early protocols and programs trusted their users. Telnet sends passwords in the clear. For a long time, SMTP would relay mail from anyone to anyone without restricting sender or receiver in any way. This "open relay" service was great for spammers—if you don't need a direct reply, you can lie about your source address, which makes fraud and denial of service attacks easy. The Internet protocols and the programs built upon them were designed for an honest, cooperative and well-meaning community of trusted parties. That is not remotely what today's

Internet is like, so on a variety of fronts, we are playing catch-up on information security and authentication.

Internet privacy and security are hard, as we'll discuss further in the following chapters. It feels like an arms race between attackers and defenders, with the attackers more often on the winning side. Data passes through shared, unregulated, and diverse media and sites scattered over the whole world, and it can be logged, inspected and impeded at any point along the path for government, commercial, and criminal purposes. It's hard to control access and to protect information along the way. Many networking technologies use broadcast, which is vulnerable to eavesdropping. Attacks on wired Ethernets and fiber optics require finding a cable and making a physical connection, but attacks on wireless don't need physical access to snoop, just proximity.

At a broader level, the overall structure and openness of the Internet is vulnerable to government control by country firewalls that block or restrict information flow in and out. There is increasing pressure on Internet governance as well, a danger that bureaucratic controls might trump technical considerations. The more that these are imposed, the greater the risk that the universal network will become balkanized and thus ultimately much less valuable.

10

The World Wide Web

"The WorldWideWeb (W3) is a wide-area hypermedia information retrieval initiative aiming to give universal access to a large universe of documents."

From the first web page, at info.cern.ch/hypertext/WWW/TheProject.html, 1990.

The most visible face of the Internet is the World Wide Web or, by now, just "the web." There's a tendency to conflate the Internet and the web, but they're different. As we saw in Chapter 9, the Internet is a communications infrastructure or substrate that lets millions of computers worldwide easily exchange information with each other. The web connects computers that provide information—servers—with computers that ask for it—clients like you and me. The web *uses* the Internet to make the connections and carry the information and it provides an interface for accessing other Internet-enabled services.

Like many great ideas, the web is fundamentally simple. Only four things matter, given the existence of a pervasive, efficient, open and basically free underlying network, which is a big proviso.

First is the *URL*, or Uniform Resource Locator, which specifies a name for a source of information, like `http://www.amazon.com`.

Second is *HTTP*, the Hypertext Transfer Protocol, which was mentioned briefly in the last chapter as an example of a higher-level protocol. An HTTP client makes a request for a specific URL and the server returns the information that was asked for.

Third is *HTML*, the Hypertext Markup Language, a language for describing the format or presentation of information returned by a server. Again, it's simple and you need know very little to make basic use of it.

Finally, there's a *browser*, a program like Chrome, Firefox, Safari or Edge that runs on your computer, makes requests of servers using URLs and HTTP, retrieves the HTML sent by a server, and displays it.

The web began life in 1989, when Tim Berners-Lee, an English computer scientist working at CERN, the European physics research center near Geneva, created a system to make scientific literature and research results more accessible over the

Internet. His design included URLs, HTTP and HTML, and there was a text-only client program for viewing what was available. There's a simulation of that first version on CERN's web site at line-mode.cern.ch/www/hypertext/WWW/TheProject.html.

This program was in use by 1990. I saw it in action in October 1992 during a visit to Cornell. It's embarrassing to admit, but at the time I didn't find it all that impressive, and I certainly didn't know that in less than six months, the creation of the first graphical browser would change the world. So much for being able to see the future.

That first browser, Mosaic, was created by students at the University of Illinois. It took off rapidly after its initial release in February 1993, and the first commercial browser, Netscape Navigator, became available only a year later. Netscape Navigator was an early success and the surge of interest in the Internet caught Microsoft unaware. The company did wake up and quickly produced a competitor, Internet Explorer (IE), which became the most widely used browser by a large margin.

Microsoft's domination of the PC marketplace raised antitrust concerns in several areas, and in 1998 the company was sued by the US Department of Justice. IE was a part of that legal proceeding because it was alleged that Microsoft was using its dominant position to drive Netscape out of business. Microsoft lost the case and was forced to alter some of its business practices.

Today Chrome is the most widely used browser on laptops, desktops and phones; Safari and Firefox are significantly less popular. In 2015, Microsoft released a new Windows 10 browser, called Edge, to replace IE. Edge originally used Microsoft's own code, but since 2019 it has been based on Google's open-source Chromium browser. Edge's market share is lower than Firefox's, and IE is even lower.

The technical evolution of the web is managed, or at least guided, by the World Wide Web Consortium, or W3C (`w3.org`), a non-profit enterprise. Berners-Lee, the founder and current director of W3C, made no attempt to profit from his invention, generously preferring to make it free for everyone, though many who jumped on the Internet and web bandwagon enabled by his work became very rich. He was knighted by Queen Elizabeth II in 2004.

10.1 How the Web Works

Let's take a more careful look at the technical components and mechanisms of the web, starting with URLs and HTTP.

Imagine viewing a simple web page with your favorite browser. Some of the text on that page is likely to be in blue and underlined; if you click on that text, the current page will be replaced by the new page that the blue text links to. Linking pages like this is called *hypertext* ("more than text"); it's an old idea but the browser has made it part of everyone's experience.

Suppose the link says something like "W3C home page." When you move the mouse over the link, the status bar at the bottom of your browser window is likely to display the URL that the link points to, something like `http://w3.org`, perhaps with further information after the domain name.

When you click on the link, the browser opens a TCP/IP connection to port 80 at the domain `w3.org` and sends an HTTP request for the information given by the rest of the URL. If the link is `http://w3.org/index.html`, the request is for the file `index.html`.

When it receives this request, the server at `w3.org` decides what to do. If the request is for an existing file on the server, the server sends that file back, and the client, your browser, displays it. The text that comes back from the server is almost always in HTML, a form that combines the real content with information about how to format or display it.

In real life, it can be this simple, but there is usually more. The protocol allows the browser to send a few lines of additional information along with the client request, and the reply from the server will generally include extra lines that indicate how much data follows and what kind it is.

The URL itself encodes information. The first part, `http`, is one of several possibilities that tell which specific protocol to use. HTTP is most common but you will see others as well, including `file` for information from the local machine (rather than from the web) and increasingly often, `https` for a secure (encrypted) version of HTTP, which we'll talk about shortly.

After `://` comes the domain name, which names the server. After the domain name there can be a slash (`/`) and a string of characters. That string is passed verbatim to the server, which can do whatever it likes with it. In the simplest case, there's nothing at all, not even the slash, and the server returns a default page like `index.html`. If there's a file name, its contents are returned as is. A question mark after the initial part of the file name usually means that the server is supposed to run a program whose name is the part before the question mark, and pass the rest of the text to that program. That's one way that information from forms on a web page is processed; for example, a Bing search is

```
https://www.bing.com/search?q=funny+cat+pictures
```

which you can confirm by typing this directly into the address bar of a browser.

The text after the domain name is written in a restricted character set that excludes spaces and most non-alphanumeric characters, so these must be encoded. A plus sign "+" encodes a space, and other characters are encoded as a `%` sign and two hex digits. For example, the URL fragment `5%2710%22%2D6%273%22` means `5'10"-6'3"`, since hex `27` is a single quote character, hex `22` is a double quote character, and hex `2D` is a minus sign.

10.2 HTML

The server response is usually in HTML, a combination of content and formatting information. HTML is so simple that it's easy to create web pages with your favorite text editor. (If you use a word processor like Microsoft Word, you must save web pages in plain text, not the default format, and with a suffix like `html`.) The formatting information is given by *tags* that describe the content and mark the beginning and often the end of regions of the page.

The HTML for a minimal web page might look like Figure 10.1. It will be displayed by a browser as shown in Figure 10.2.

```
<html>
  <title> My Page </title>
  <body>
    <h2> A heading </h2>
    <p> A paragraph... </p>
    <p> Another paragraph ... </p>
      <img src="wikipedia.jpg" alt="Wikipedia logo" />
      <a href="http://www.wikipedia.org">link to Wikipedia</a>
      <h3> A sub-heading </h3>
        <p> Yet another paragraph </p>
  </body>
</html>
```

Figure 10.1: HTML for a simple web page.

Figure 10.2: Browser display of HTML from Figure 10.1.

The image file by default comes from the same place as the original file but it could come from anywhere on the web. If the file named in the `<img>` image tag is not accessible, the browser will display some "broken" image in its place. The `alt=` attribute provides text to be displayed if the image itself can't be shown; it's one small example of a web-page technique that can help people who might have vision or hearing problems.

Some tags are self-contained, like `<img>`; some have a beginning and end, like `<body>` and `</body>`. Others, like `<p>`, do not need a closing tag in practice, though a strict definition does require `</p>`, which we have used here. Indentation and line breaks are not necessary but make the text easier to read.

Most HTML documents also contain information in another language called CSS (Cascading Style Sheets). With CSS one can define style properties like the format of headings in a single place and have them apply to all occurrences. For example, we could cause all h2 and h3 headings to display in red italics with this CSS:

```
h2, h3 { color: red; font-style: italic; }
```

Both HTML and CSS are *languages*, but not *programming languages*. They have formal grammars and semantics but they don't have loops and conditionals, so you can't express an algorithm in them.

The point of this section is to show just enough of HTML to demystify how web pages work. Considerable skill is required to create the polished web pages that you see on commercial sites, but the basics are so simple that with a few minutes of study you can make your own more modest pages. A dozen tags will see you through the creation of most text-only web pages and another dozen are enough to do pretty much anything that a casual user might care about. It's easy to create pages by hand, word processors have a "create HTML" option, and there are programs specifically aimed at creating professional-looking web pages. You will need such tools if you're going to do serious web design, but it's always good to understand how things work underneath.

The original design of HTML only handled plain text for the browser to display. It wasn't long, however, before browsers added the capability to display images, including simple artwork like logos and smiley faces in GIF format and pictures in JPEG format. Web pages offered forms to be filled in, buttons to push, and new windows that popped up or replaced the current one. Sound, animations, and movies followed soon after, generally once there was enough bandwidth to download them quickly and enough processing power to display them.

There is also a simple mechanism, the unintuitively named CGI or Common Gateway Interface, for passing information from the client (your browser) to a server, for example a name and password, or a search query, or the choices made with radio buttons and dropdown menus. This mechanism is provided by the HTML <form> ... </form> tag. Within a <form>, you can include common user interface elements like text entry areas, buttons, checkboxes, and so on. If there is a "Submit" button, pushing it causes the data within the form to be sent to the server, along with a request to run a specific program using that data.

Forms have limitations. They support only a few kinds of interface elements. Form data can't be validated except by writing JavaScript code or sending it to the server for processing. There is a password input field that replaces typed characters by asterisks, but it provides no security whatsoever, since the password is transmitted and stored in logs without encryption. Nevertheless, forms are a crucial part of the web.

10.3 Cookies

The HTTP protocol is *stateless*, a bit of jargon which means that an HTTP server is not required to remember anything about client requests—it can discard all records of each exchange after it returns the requested page.

Suppose that the server really does need to remember something, perhaps the fact that you have already provided a name and password so there's no need to keep asking for them on subsequent interactions. How might that be made to work? The problem is that the interval between the first visit and the second might be hours or weeks or might never happen at all, and that's a long time for a server to retain information on speculation.

In 1994, Netscape invented a solution that it called a *cookie*, a cutesy but well-established programmer word for a small piece of information passed between programs. When a server sends a web page to a browser, it can include additional chunks of text (up to about 4,000 bytes each) that the browser is meant to store; each chunk is called a cookie. When the browser makes a subsequent request to the same server, it sends those cookies back. In effect the server uses memory on the client to remember something about the client's previous visit. Often the server will assign a unique identification number to a client and include that in a cookie; permanent information associated with that identification number is maintained in a database on the server. This might be login status, shopping cart contents, user preferences, and the like. Each time the user revisits the site, the server can use the cookie to identify him or her as someone seen before and set up or restore information.

I normally disallow all cookies, so when I visit Amazon, the initial page greets me with "Hello." But if I want to buy something, I have to log in and add it to my shopping cart, and that requires me to allow Amazon to set cookies. Thereafter, each visit says "Hello, Brian" until I delete those cookies.

Each cookie has a name, and a single server may store multiple cookies with each visit. A cookie is not a program and it has no active content. Cookies are entirely passive: they're just strings of characters that are stored and subsequently sent back; nothing goes back to the server that didn't originate with the server. They are only sent back to the domain from which they originated. Cookies have an expiration date after which they are deleted by the browser. There's no requirement that the browser accept or return them.

It's easy to view the cookies on your computer; the browser itself will show them to you, or you can use other tools. For example, a recent visit to Amazon deposited half a dozen cookies. Figure 10.3 shows them as seen through Cookie Quick Manager, a Firefox extension. Notice that Amazon appears to have detected that I am using an ad blocker.

Figure 10.3: Cookies from Amazon.

In principle, this all sounds benign and it certainly was intended as such, but no good deed goes unpunished and cookies have been co-opted into less desirable uses as well. The most common is to track people as they browse, to create a record of the sites they have visited and then to provide targeted advertisements. We'll talk about how this works in the next chapter, along with other techniques that track you as you travel around the web.

10.4 Active Content in Web Pages

The original design of the web took no particular advantage of the fact that the client is a powerful computer, a general-purpose programmable device. The first browsers could make server requests on behalf of the user, send information from forms and, with the aid of helper programs, display content like pictures and sound that required special processing. But browsers soon made it possible to download code from the web and execute it; this is sometimes called *active content.* As might be expected, active content has significant consequences, some good and some definitely not.

Early versions of Netscape Navigator included a way to run Java programs within the browser. At the time Java was a comparatively new language. It had been designed to be installed in environments with modest computing capabilities (like home appliances), so it was technically feasible to include a Java interpreter along with the browser. This offered the prospect of being able to do significant computation in the browser, perhaps replacing conventional programs like word processors and spreadsheets, and even the operating system itself. The idea was sufficiently worrisome to Microsoft that it took a series of actions to undermine the use of Java. In 1997, Sun Microsystems, the creator of Java, sued Microsoft, a legal process that was settled some years later with Microsoft paying Sun well over a billion dollars.

For a variety of reasons, Java never took off as the way to extend browsers. Java itself is widely used but its integration with the browser is limited and today it is rarely used in that role.

Netscape created a new language specifically for use within its browsers: JavaScript, which appeared in 1995. In spite of its name, which was chosen for marketing reasons, JavaScript is unrelated to Java except that both have a superficial similarity to the C programming language, as we saw in Chapter 5. Both use a virtual machine implementation, but with significant technical differences. Java source code is compiled where it is created and the resulting *object code* is sent to the browser for interpretation; you can't see what the original Java source code looks like. By contrast, JavaScript *source code* is sent to the browser and is compiled there. The recipient can see the source code that is being executed, and could study and adapt it as well as run it.

Almost all web pages today include some JavaScript, to provide graphical effects, validate information in forms, pop up both useful and irritating windows, and so on. Its use for advertising popups has been mitigated by popup blockers, now included with browsers, but its use for sophisticated tracking and monitoring is pervasive. JavaScript is so ubiquitous that it's hard to use the web without it, though browser add-ons like NoScript and Ghostery give us some control over what JavaScript code will be run. Somewhat ironically, add-ons are themselves written in JavaScript.

On balance, JavaScript is more good than evil, though there are days when I might come down on the other side, especially considering how much it's used for tracking (which we will discuss in Chapter 11). I routinely disable JavaScript entirely with NoScript, but then have to restore it selectively for sites that I care about.

Other languages and contents are also handled by the browser, either with code in the browser itself or by *plug-ins* like Apple QuickTime and Adobe Flash. Plug-ins are programs, usually written by third parties, that are dynamically loaded into the browser as needed. If you visit a page that has content in a format that your browser can't already handle, you may be offered the chance to "get the plug-in." What that means is that you will be downloading a new program to run on your computer in close cooperation with the browser.

What can a plug-in do? Essentially anything it wants to, so you are pretty much forced to trust the supplier, or do without the content. A plug-in is compiled code that executes as part of the browser using an API presented by the browser and in effect it becomes part of the browser when it runs. Flash is widely used for video and animations; Adobe Reader for PDF documents is another common plug-in. The short story on plug-ins is that if you trust their source, you can use them with no more than the normal perils of code that will have bugs and can monitor your behavior. Flash has had a long history of major security vulnerabilities, however, and its use is now deprecated. HTML5 provides browser facilities that reduce the need for plug-ins, especially for video and graphics, but plug-ins are likely to be around for a long time.

As we saw in Chapter 6, browsers are like specialized operating systems, which can be extended to handle richer and ever more complicated content "to enhance your browsing experience." The good news is that it's possible to do a lot with a program running in the browser, and interactions will run faster if computation is performed locally. The downside is that this requires your browser to execute programs that someone else wrote and whose properties you almost surely don't understand. There are real risks in running code from unknown sources on your computer. "I have always depended on the kindness of strangers" is not a prudent security policy. In a Microsoft article titled *10 Immutable Laws of Security*, the first law is "If a bad guy can persuade you to run his program on your computer, it's not your computer any more." Be conservative in allowing JavaScript and plug-ins.

10.5 Active Content Elsewhere

Active content can appear in places other than web pages. Consider email. Once mail has arrived, it will be displayed by a mail-reading program. Obviously a mail reader has to display text; the question is how far it goes in interpreting other kinds of content that might be included, since that has a significant effect on privacy and security.

What about HTML in mail messages? Interpreting size and font tags is benign: there's no risk in displaying part of a message in big red letters, though it might irritate the recipient. Should the mail reader display images automatically? That makes it easy to view pictures, but opens up the prospect of more cookies if content comes from other sources. We could block email cookies, but there's nothing to prevent a mail sender from including an image consisting of a 1-by-1 transparent pixel whose URL encodes something about the message or the recipient. (These invisible images are sometimes called *web beacons*; they are frequent on web pages.) When your HTML-enabled mail reader requests the image, the site that serves it knows that you are reading that specific mail message at a particular time. This provides an easy way to track when mail has been read and potentially reveals information that you might prefer to keep private.

What happens if a mail message includes JavaScript? What if it includes a Word or Excel or PowerPoint document? Should the mail reader automatically run those programs? Should it make it easy for you to do so by clicking somewhere in the message? Come to think of it, should it let you click directly on links in messages? That's a favorite way to entice victims to do something foolish. PDF documents can include JavaScript (which surprised me when I first saw it); should that code be executed automatically by the PDF viewer that is invoked automatically by the mail reader?

It's convenient to attach documents, spreadsheets and slide presentations to email messages and it's standard operating procedure in most environments, but such documents can carry viruses, as we shall see shortly, and blindly clicking on them is one way to propagate viruses.

It's even worse if mail messages include executable files, like Windows `.exe` files or equivalents. Clicking on one of those launches the program, which with high probability is something that will damage you or your system. The bad guys use a variety of tricks to get you to run such programs. I once got mail that claimed to include a picture of Anna Kournikova, the Russian tennis player, and encouraged me to click on it. The file name was `kournikova.jpg.vbs`, but the `.vbs` extension was hidden (a misguided feature of Windows), concealing the fact that it was not a picture but a Visual Basic program. Fortunately, since I was using an antiquated text-only mail program on a Unix system, clicking was not an option, so I saved the "picture" in a file for later inspection.

10.6 Viruses, Worms and Trojan Horses

The Anna Kournikova "picture" was in fact a virus. Let's talk a bit about viruses and worms. Both of these terms refer to (often malicious) code that propagates from one system to another. The technical distinction, not very important, is that a virus needs help with propagation—it can only get to another system if you do something to enable it—while a worm can propagate without your assistance.

Although the potential for such programs had been known for a long time, the first example that made the evening news was the "Internet worm" launched by Robert T. Morris in November 1988, well before what we would call the modern Internet era. Morris's worm used two different mechanisms to copy itself from one system to another, relying on bugs in widely used programs, together with a dictionary attack (trying common words as potential passwords) so it could log itself in.

Morris had no malicious intent; he was a computer science graduate student at Cornell and had been planning an experiment to measure the size of the Internet. Unfortunately a programming error made the worm propagate much faster than he expected, with the result that many machines were infected multiple times, couldn't cope with the volume of traffic, and had to be disconnected from the Internet. Morris was convicted of a felony under the then-new Computer Fraud and Abuse Act, and was required to pay a fine and perform public service.

For some years, it was common for viruses to propagate via infected floppy disks, which before the widespread use of the Internet were a standard medium for exchanging programs and data between PCs. An infected floppy included a program that would run automatically when the floppy was loaded; the program would copy itself to the local machine in such a way that the virus would be passed on whenever a subsequent floppy disk was written.

Virus propagation became much easier with the arrival in 1991 of Visual Basic in Microsoft Office programs, especially Word. Most versions of Word include a VB interpreter and Word documents (`.doc` files) can contain VB programs, as can Excel and PowerPoint files, among others. It is dead simple to write a VB program that will take control when a document is opened, and because VB provides access to the entire Windows operating system, the program can do anything it wants. The usual sequence is for the virus to install itself locally if it's not already present, and then arrange to propagate itself to other systems. In one common mode of propagation, when an infected document is opened, the virus mails a copy of itself, along with some innocuous or appealing message, to every entry in the current victim's email address book. (The Anna Kournikova virus used this method.) If the recipient opens the document, the virus installs itself on the new system and the process repeats.

In the mid to late 1990s, there were many such VB viruses. Because the default setting for Word at the time was to blindly run VB programs without asking for permission, infections spread rapidly and large corporations had to turn off all their computers and clean them up individually to stamp out the virus. VB viruses are still around but merely changing the default behavior of Word and similar programs has greatly reduced their impact. In addition, most mail systems now strip out VB programs and other suspicious content from incoming mail before the mail reaches a human reader.

VB viruses are so easy to create that they could even be written by inexperienced programmers, who were called "script kiddies." Making a virus or worm that does its job without getting caught is harder. In late 2010, a sophisticated worm called Stuxnet was discovered in a number of process control computers. Its main target was uranium enrichment equipment in Iran. Its approach was subtle: it caused speed fluctuations in centrifuge motors that would lead to damage or even destruction that might seem like normal wear and tear; at the same time, it told monitoring systems that everything was fine so nobody would notice problems. No one has stepped forward to take credit for this program, though it is widely believed to have involved Israel and the US.

A *Trojan horse* (often shortened to Trojan in this context) is a program that masquerades as something beneficial or harmless but in fact does something harmful. The victim is induced to download or install the Trojan because it appears to do something helpful. One typical example offers to perform a security analysis on a system, but in fact installs malware instead.

Most Trojan horses arrive by email. The message in Figure 10.4 (lightly edited) has a Word attachment that, if incautiously opened in Windows, installs malware known as Dridex. This attack was easy to spot, of course—I don't know the sender, I have never heard of the company, and the sender address is unrelated to that company. Even if I hadn't been alert, I use a text-only mail program on Linux so I'm pretty safe; this attack is targeted at Windows users. (I've since received at least two dozen variants of this message, of varying plausibility.)

```
From: Efrain Bradley <BradleyEfrain90@renatohairstyling.nl>
Subject: Invoice 66858635 19/12
Hi,
Happy New Year to you !  Hope you had a lovely break.
Many thanks for the payment. There's just one invoice that hasn't
been paid and doesn't seem to have a query against it either.
Its invoice  66858635  19/12  ?4024.80  P/O ETCPO 35094
Can you have a look at it for me please?  Thank-you !
Kind regards
Efrain Bradley
Credit Control, Finance Department, Ibstock Group
Supporting Ibstock, Ibstock-Kevington & Forticrete
---------------------------------------------
( +44 (0)1530 dddddddd
[ Attachment: "invoice66858635.doc" 18 KB. ]
```

Figure 10.4: A Trojan horse attempt.

I mentioned floppy disks as an early vector for propagating viruses. One modern equivalent is an infected USB flash drive. You might think that a flash drive is a passive device, since it's just memory. However, some systems, notably Windows, provide an "autorun" service that automatically runs a program *from the drive* when a CD, DVD or flash drive is plugged in. If this feature is enabled, malevolent software can be installed and damage done without any warning or chance to intervene. It's not uncommon for corporate systems to be infected in this fashion, even though most companies have strict policies against plugging USB drives into company computers. On occasion, brand new drives have been shipped with viruses already on them, a kind of "supply chain" attack. An easier attack is to leave a drive with a company's logo on it in the company parking lot. If the drive contains a file with an intriguing name like "ExecutiveSalaries.xls," autorun may not even be needed.

10.7 Web Security

The web raises many difficult security issues. Broadly speaking, one might divide the threats into three categories: attacks on clients (that's you), attacks on servers (online stores or your bank, for example), and attacks on information in transit (like snooping on your wireless or the NSA grabbing all the traffic on a fiber optic cable). Let's talk about each in turn, to see what might go wrong and what can be done to alleviate the situation.

10.7.1 Attacks on clients

Attacks on you include nuisances like spam and tracking, as well as more serious concerns, especially the release of private information like your credit card and bank account numbers or passwords that could let someone else masquerade as you.

In the next chapter, we will discuss in detail how cookies and other tracking mechanisms are used to monitor your web activities, ostensibly to provide you with more relevant and thus less irritating advertising. Tracking can be reduced by forbidding *third-party cookies* (that is, cookies that come from a different web site than the one you went to), and by using browser add-ons that disable trackers, turn off JavaScript, and so on. It's a nuisance to maintain your defenses, since many sites are unusable when you have your shields all the way up—you have to lower them temporarily, then remember to reset—but I think it's worth the trouble. Browser suppliers are making it easier to block some cookies and other trackers, though you might have to adjust default settings, and external blockers are still worthwhile.

Spam—unsolicited mail that offers get-rich schemes, stock tips, body-part enhancements, performance improvers, and a host of other unwanted goods and services—has become so voluminous as to endanger the very use of email. I generally get fifty to a hundred spam messages per day, more than the number of real messages. Spam is so common because sending it is almost free. If even the tiniest fraction of the millions of recipients reply, that's enough to keep things profitable.

Spam filters try to separate the wheat from the chaff by analyzing the text for known patterns ("Tasty drink wipes out unwanted extra fat," promises a printable one from a recent batch), unlikely names, bizarre spellings (\/l/-\GR/-\), or addresses that spammers favor. No single criterion could be sufficient, so a combination of filters is used. Spam filtering is a major application of *machine learning*. Given a training set of examples that have been tagged as spam or not spam, a machine learning algorithm classifies subsequent inputs based on their similarity to the characteristics of the training set. We will have more to say about this in Chapter 12.

Spam is an example of an arms race, since as the defenders learn how to cope with one kind of spam, the offenders find a new way around. Spam is hard to stop at the source, because its source is well hidden. Much spam is sent by compromised personal computers, often running Windows. Security holes and lax management by users leave the computers vulnerable to the installation of *malware*, that is, malicious software that intends to do damage or otherwise interfere with the system. One kind of malware will broadcast spam mail upon command from an upstream control, which might in turn be controlled from another computer; each step makes it that much harder to find the originator.

Phishing attacks try to convince the recipient to voluntarily hand over information that can be used for theft. You've almost surely gotten a "Nigerian" scam letter. (Oddly, the last several to me have been in French, as in Figure 10.5.) It's hard to believe that anyone ever responded to something so implausible, though apparently people still do.

Most phishing attacks are more subtle. Plausible mail arrives, ostensibly from a legitimate institution or a friend or co-worker, asking you to visit a site or read a document or verify some credentials. If you do, your adversary now has installed something on your computer or he has information about you; in either case, he can potentially steal your money or your identity, or attack your employer. Fortunately, grammar and spelling mistakes may give it away, and mousing over links often reveals that they lead to somewhere suspicious.

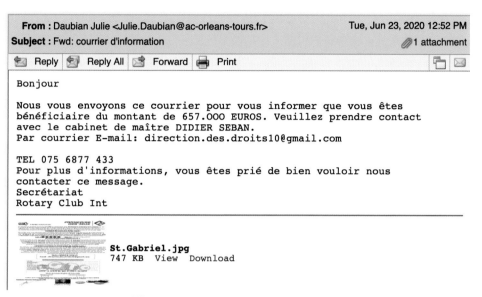

Figure 10.5: Phishing from France.

It's easy to create a message that looks official since format and images like a company logo can be copied from the real site. The return address doesn't matter because there's no check that the sender is legitimate. Like spam, phishing costs almost nothing, so even a tiny success rate can be profitable.

Figure 10.6 is an edited transcript of a somewhat targeted attack that began with mail, ostensibly from a colleague who I will call JP. Since I had seen a similar attack a few weeks earlier, and the actual mail address was deceptive, a variation on jp.princeton.edu@gmail.com, I decided to play along.

```
JP: Are you available for a quick task?

BK: what's up?

JP: Okay, I'm in a meeting, i need ebay gifts card
purchased, let me know if you can quickly stop by the
nearest store so i can advise the quantity and the
denominations to procure.  Turn in the expense for
reimbursement later.
Thanks

BK: what kind of store?  nearest one is a liquor store.

JP: Okay, Pick up 5 ebay gifts card at $200/each = $1000.You
can get the at any store around around Scratch-off silver
lining at the back for the pin codes.  Send the pin codes on
each cards once purchased.  Can you go take care of this
now?

BK: I don't think the liquor store has that kind of card,
and I normally just buy some beer.  Any other suggestions?
```

Figure 10.6: A weak phishing attempt.

The perps eventually gave up; they must have been on to me. This attack was marginally convincing because it was targeted at me and used the name of a colleague and friend. It's for this reason that such precisely targeted attacks are sometimes called *spear phishing*. Spear phishing is a kind of *social engineering*: induce the victim to do something foolish by pretending to have a personal relationship like a mutual friend, or claiming to work for the same company. The more of your life that you reveal in places like Facebook and LinkedIn, the easier it is for someone to target you. Social networks help social engineers.

In July 2020, Twitter suffered an embarrassing attack in which the accounts of a number of high-profile individuals, like Bill Gates, Jeff Bezos, Elon Musk, Barack Obama and Joe Biden, were used to tweet variations on "Send us $1,000 in Bitcoin and we will send you back $2,000." It's hard to believe that anyone would fall for this, let alone someone capable of sending bitcoins, but apparently hundreds of people did before Twitter managed to shut it down. Twitter subsequently said

> The social engineering that occurred on July 15, 2020, targeted a small number of employees through a phone spear phishing attack. A successful attack required the attackers to obtain access to both our internal network as well as specific employee credentials that granted them access to our internal support tools. Not all of the employees that were initially targeted had permissions to use account management tools, but the attackers used their credentials to access our internal systems and gain information about our processes. This knowledge then enabled them to target additional employees who did have access to our account support tools.

Notice that the attackers were able to escalate from employees without sufficient access to ones who did have access.

The primary person behind the attack was quickly identified, a 17-year-old from Florida; two other young men were also charged.

Spear phishing or social engineering attacks that ostensibly come from the CEO or other high-level executives seem especially effective. One popular version in the months before income tax returns are due asks the target to send tax information on each employee, like the W-2 form in the US. That form includes an accurate name, address, salary and Social Security number, so it can be used to file for a fraudulent tax refund. By the time the employees and the tax authorities notice, the perpetrators have the money and are long gone.

Spyware refers to programs that run on your computer, sending information about you to somewhere else. Some of this is clearly malevolent, though sometimes it's merely commercial snooping. For instance, most current operating systems check automatically for updated versions of installed software. One could argue that this is a good thing, since it encourages you to update software to get bug fixes for security problems, but it could equally well be called an invasion of privacy—it's no one's business what software you're running. If the updates are forced on you, that can be a problem: in too many cases, newer versions of programs are bigger but not necessarily better, and a new version can break existing behavior or add new bugs. I try to avoid updating critical software during a semester because it might change something I need for a class.

On personal machines, it's common for an attacker to install a *zombie*, that is, a program that waits until it is awakened via the Internet and told to perform some hostile act like sending spam. Such programs are often called *bots* and networks of them with a common control are called *botnets*. At any point, there are thousands of known botnets and millions of bots that can be called into service. Selling bots to potential attackers is a thriving business.

One can compromise a client computer and steal information at the source, either by looking for it in the file system or by using a surreptitiously installed *key logger* to capture passwords and other data as they are entered. A key logger is a program that monitors all keystrokes on the client and thus is able to capture passwords as they are typed; encryption cannot help here. Malware could also turn on the microphone and camera in the computer.

It's possible for malware to encrypt the contents of your computer so you can't use it until you pay for a decryption password; naturally this kind of attack is called *ransomware*. In June 2020, the medical school at the University of California, San Francisco (UCSF) was hit by an attack; in a statement, the university said

> The data that was encrypted is important to some of the academic work we pursue as a university serving the public good. We therefore made the difficult decision to pay some portion of the ransom, approximately $1.14 million, to the individuals behind the malware attack in exchange for a tool to unlock the encrypted data and the return of the data they obtained.

Not long after this, I received mail from a scientific institution that I belong to, reporting on a ransomware attack that might well have included data about me. The institution uses a company called Blackbaud to provide services (think cloud computing). Here's part of the mail:

> We were also informed by Blackbaud that in order to protect data and mitigate potential identity theft, it met the cybercriminal's ransomware demand. Blackbaud has advised us that it received assurances from the cybercriminal and third-party experts that the data was destroyed.

> We are continuing to work with Blackbaud to understand why there was a delay between it finding the breach and notifying us...

The "delay" was about two months, during which time Blackbaud paid up; the institution in question delayed another two weeks before informing me and presumably other members. I also wonder about the "assurances" from the bad guys that they have destroyed the information. Does that remind you of blackmailers who promise to destroy the incriminating photographs?

A simpler version of ransomware just pops up a threatening screen that claims your computer is infected with malware, but you can get rid of it: don't touch anything but call this toll-free number and for a modest fee, you'll be rescued. This is a kind of *scareware*. A relative of mine fell for this fraud and paid several hundred dollars. Fortunately the credit card company reversed the charges; not everyone is so lucky. If you pay with bitcoins, you have no recourse at all if the bad guys don't honor the agreement.

The risks are exacerbated if your browser or other software has bugs that make it possible for miscreants to install their software on your machine. Browsers are large, complicated programs and they have had numerous bugs that permit attacks on users. Keeping your browser up to date is a partial defense, as is configuring it so it does not release unnecessary information or permit random downloading. For example, set your browser preferences so the browser will ask for confirmation before opening content types like Word or Excel documents. Be cautious about what you download; never just click when a web page or program asks you to do so. We'll discuss more defenses in a few pages.

On phones, the most likely risks are downloading apps that will export your personal information. An app can access all information on the phone, including contacts, location data and call records, and can easily use that against you. Phone software is slowly getting better at helping you defend yourself, for example by giving you more fine-grained control over permissions, but the emphasis is still on "slowly."

10.7.2 Attacks on servers

Attacks on servers aren't your problem directly, in that there isn't much you can do about them, but that doesn't mean that you can't be victimized.

Servers have to be carefully programmed and configured so that client requests, no matter how cleverly crafted, cannot cause the server to release unauthorized information or permit unauthorized access. Servers run big complicated programs, so bugs and configuration errors are common; both can be exploited. Servers are usually supported by databases that are accessed through a standard interface called SQL ("structured query language"). One frequent attack is called *SQL injection*. If access is not carefully circumscribed, a clever attacker can submit queries that reveal the database structure, extract unauthorized information, and even run the attacker's code on the server, code that might be able to gain control of the whole system. Such attacks are well understood, as are defenses against them, but they still occur surprisingly often.

Once a system has been compromised, there are few limits on the harm that can be done, especially if the attacker has managed to obtain "root" access, that is, access at the highest level of administrative privilege. This is true whether the target is a server or an individual home computer. At that point, an attacker can vandalize or deface a web site, post embarrassing material like hate speech, download destructive programs, or store and distribute illegal content like child pornography and pirated software. Data can be stolen in bulk from servers or in modest quantities from individual machines.

Such breaches are now an almost daily event, and sometimes on a grand scale. In March 2017, terabytes of personally identifiable information on 150 million people were copied from Equifax, one of three credit reporting companies in the US. Credit agencies like Equifax hold a great deal of sensitive information in their databases, so this was potentially a serious problem. Equifax was derelict in their security procedures—they had not kept their systems up to date against known vulnerabilities—and their behavior after the breach wasn't all that good either. The company did not publicly reveal the breach until September, and some high-level executives sold stock before the breach was made public.

In December 2019, Wawa, a US chain of convenience stores, announced that information from a large number of credit cards, perhaps 30 million, had been stolen via malware that found its way onto Wawa sales terminals; the card information was for sale on the dark web.

In February 2020, Clearview AI, a company that provides facial recognition software primarily though not exclusively to law enforcement agencies, was broken into, and its client database was leaked. The company claimed that nothing else was stolen, including photographs and records of searches made, though news stories at the time implied that the photos were also taken.

Also in February 2020, the Marriott International hotel chain announced that information on over 5 million guests had been stolen; the information included contact details and other personal facts like date of birth.

Servers can also be subject to *denial of service* attacks, in which an adversary directs a large amount of traffic at a site to overwhelm it with sheer volume. This is often orchestrated with botnets; the compromised machines are told to send requests to a particular site at a particular time, leading to a coordinated flood. An attack that comes from many sources simultaneously is called a *distributed denial of service* (DDoS) attack. As an example, in February 2020, Amazon's AWS cloud service successfully dealt with what it says is the largest DDoS attack ever, with a peak traffic rate of 2.3 Tbps.

Although denial of service attacks are most often large and aimed at big servers, small-scale versions are also possible. For instance, my employer recently replaced a convenient home-grown appointment-scheduling system with a commercial offering that accesses a user's online calendar to find and populate open time slots. The company calls it "painless scheduling." Go to a web link with the user's identification, click on an open slot, provide a confirmation email address, and you're done. But there is *no* checking of anything, so if I can guess your identification, I can anonymously fill up all your available slots. There's no validation of the email addresses either, so I can use the calendar system as a vector for sending anonymous nuisance messages to anyone. I'd be seriously disappointed if one of my student project groups created such a privacy and security hole; one would expect better from a pricey commercial product.

10.7.3 Attacks on information in transit

Attacks on information in transit have perhaps been the least of the concerns, though they are certainly still serious and common enough; with the spread of wireless systems, this may change, and not for the better. Someone who wants to steal money could eavesdrop on your conversation with your bank to collect your account number and password. If traffic between you and the bank is encrypted, however, it can't be understood. Programs can snoop on unencrypted connections anywhere that offers open wireless access and may allow an attacker to pretend to be you, quite undetectably. One large theft of credit card data involved listening to unencrypted wireless communications between terminals in stores; the thieves parked outside the stores and captured credit card information as it went by.

HTTPS is a version of HTTP that encrypts TCP/IP traffic in both directions, which makes it impossible for an eavesdropper to read contents or to masquerade as one of the parties. HTTPS use is growing quickly, though it is not yet universal.

It's also possible to mount a *man-in-the-middle* attack, in which the attacker intercepts messages, modifies them, and sends them along as if they had come directly from their original source. (The *Count of Monte Cristo* story mentioned in Chapter 8 is an instance.) Proper encryption prevents this kind of attack as well. Country firewalls are another kind of man-in-the-middle attack, where traffic is slowed or search results are altered.

A *virtual private network* (VPN) establishes an encrypted path between two computers, and thus generally secures information flow in both directions. Corporations often use VPNs to enable employees to work from home or in countries where the security of communication networks can't be trusted. Individuals can use VPNs to work more safely from coffee shops and other sites that offer open Wi-Fi. But beware of who runs the VPN and how much they will stand up to government pressures to reveal information about their users.

Indeed, beware of their basic honesty and competence. In July 2020, a number of free VPN services that claimed not to log connections suffered a breach that revealed over a terabyte of logging information about their users, along with dates, times, IP addresses and even unencrypted passwords.

Secure messaging apps like Signal, WhatsApp and iMessage provide encrypted voice, video and text connections between their users. All communication is encrypted end to end, that is, it is encrypted at the source and decrypted at the receiver, using keys that exist only at the end points, not in the hands of a service provider, so in principle no one else can eavesdrop or do a man-in-the-middle attack. Facebook Messenger, another messaging app, is *not* end-to-end encrypted at the moment, though the option exists. Unless it is encrypted, it is more vulnerable to attack.

Signal is open source software, while WhatsApp is a Facebook product and iMessage comes from Apple. Edward Snowden has endorsed Signal as the preferred system for secure communications, and uses it himself.

The Zoom video conferencing system that many of us now use claimed to provide end-to-end encryption of meetings, using 256-bit AES. But a complaint filed by the US Federal Trade Commission in 2020 alleged that in fact Zoom actually retained the encryption keys, only used AES-128 encryption, and quietly installed software that bypassed a Safari browser security mechanism.

10.8 Defending Yourself

Defense is tough. You need to defend against all possible attacks, but attackers only need to find one weakness; the advantage lies with the attackers. Nevertheless, you can improve your odds, especially if you are realistic in your assessments of potential threats.

What can you do in your own defense? When someone asks me for advice, this is what I tell them. I divide defenses into three categories: ones that are very important, ones that are prudent and cautious, and ones that depend on how paranoid you are.

(Just so you know, I'm well out on the paranoid end; most people won't go so far.)

Important:

Choose passwords thoughtfully so that no one can guess them and using a computer to try a bunch of possibilities isn't likely to reveal them quickly. You need something stronger than single words, your birthday, names of family or pets or significant others, and especially variants of "password" itself, which are chosen amazingly often. A phrase of several words that includes upper and lower case letters, numbers and special characters is a decent compromise between safety and ease of use. There are a number of sites that will estimate the strength of a proposed password. The conventional wisdom is that you should change your passwords from time to time, though I'm not convinced. Frequent changes may be counter-productive, especially if forced on you at an inconvenient time, since that encourages obviously formulaic changes like incrementing a final digit.

Never use the same password for critical sites like your bank and email as you do for throwaways like online news and social media sites. Never use the same passwords at work as you do for personal accounts. Don't use a single site like Facebook or Google for signing in to other sites: it's a single point of failure if something goes wrong, and of course you're just giving away information about yourself. You can check whether a particular password has already been cracked at haveibeen-pwned.com, which collects information from breaches.

Password managers like LastPass generate and store safe random passwords for all your sites; you only have to remember one master password. Of course this is a single point of failure if you forget your password, or if the company or software holding the passwords is compromised or coerced.

Use two-factor authentication if it's available. Two-factor authentication requires both a password and a physical device in the possession of the user; it's safer than a password alone since it requires the user to know something (password) and to have something (a device). The device can be a cell phone app that generates a number which has to match a number generated at the server end by the same algorithm. It could be a message sent to your phone. Or it could be a special-purpose device like the one in Figure 10.7 that displays a freshly generated random number that you have to provide along with your password.

Figure 10.7: RSA SecurID 2-factor device.

Ironically, RSA, a company that makes a widely used two-factor authentication device called SecurID (Figure 10.7), was hacked in March 2011. Security information was stolen, rendering some SecurID devices vulnerable.

Don't open attachments from strangers and don't open unexpected attachments from friends or co-workers. Disallow Visual Basic macros in Microsoft Office programs. Never automatically accept, click or install when requested. Don't download programs of dubious provenance; be wary about downloading and installing any software, including the defensive add-ons of this section, unless it comes from a trusted source. This is just as true for your phone as it is for your computer!

Don't do anything important at places that offer open Wi-Fi—don't do your banking at Starbucks. Make sure that connections use HTTPS, but don't forget that HTTPS only encrypts the contents. Everyone on the path knows the sender and receiver; such metadata can be very useful in identifying people.

Use anti-virus software and keep it up to date. Don't click on popups that offer to run a security check on your computer. Keep software like browsers and the operating system up to date, since security fixes are frequent.

Back up your information to a safe place regularly, automatically with a service like Apple's Time Machine or manually if you are diligent. Making regular backups is a wise practice anyway, and it will make you a lot happier if a drive dies or if malware trashes a disk or encrypts it for ransom. If you use a cloud service to store precious documents and pictures, make your own backup copies too, in case you get locked out or they go out of business.

Prudent and Cautious:

Turn off third-party cookies. It's a nuisance that cookies are stored on a per-browser basis, so you have to set up defenses for each browser you use, and the details of how to enable them are different, but it's worth the effort.

Use add-ons like Adblock Plus, uBlock Origin, and Privacy Badger to reject advertising and the tracking and potential malware they enable. Use Ghostery to eliminate most JavaScript tracking. Adblock and similar add-ons work by filtering out HTTP requests to URLs in long lists of advertising sites. Advertisers claim that users of ad blockers are somehow cheating or stealing, but as long as advertisements are one of the major vectors by which malware is delivered, it's just good hygiene to disable them, and you'll discover that your browser seems faster too.

Turn on private browsing or incognito mode and remove cookies at the end of each session, though this only affects your own computer; you can still be tracked online. The Do Not Track setting doesn't do much good and can make you more identifiable.

Turn off location services on your phone unless you need a map or navigation.

Disable HTML and JavaScript in your mail reader.

Turn off operating system services that you don't use. For example, my Mac offers to let me share printers, files and devices, and to allow other computers to log in and manage my computer remotely. Windows has analogous settings. I turn them all off.

A *firewall* is a program that monitors incoming and outgoing network connections and blocks those that violate access rules. Turn on the firewall on your computer.

Use passwords to lock your phone and your laptop. If there's a fingerprint reader, use it.

Paranoid:

Use NoScript in your browser to curtail JavaScript.

Turn off all cookies except for sites that you explicitly whitelist.

Use a fake email address for temporary signups. I use mailinator.com or yop-mail.com when some site insists on an email address before they will let me access some service or information.

Turn off your phone when you're not using it. Encrypt your phone; this is automatic on newer versions of iOS and available on Android. Encrypt your laptop too.

Use the Tor browser for anonymous browsing. (More on this in Chapter 13.)

Use Signal, WhatsApp or iMessage for secure communication, but note that these may still pass along malware if you're not careful.

As cell phones become more of a target, increased precautions are necessary for them; be especially careful of apps and other content that you download. In May 2018, Jeff Bezos, founder of Amazon, appears to have had his cell phone hacked by agents of the Saudi government, via a malicious video that was included in a Whats-App message.

You can also be assured that the Internet of Things (IoT) has similar problems, for which the precautions will be harder to take because you have little control over such devices. Bruce Schneier's book *Click Here to Kill Everybody* is an excellent survey of the perils of the IoT.

10.9 Summary

The web has grown from nothing in 1990 to an essential part of our lives today. It has changed the face of business, especially at the consumer level, with search, online shopping, ratings systems, price comparison and product evaluation sites. It has changed our behavior as well, from how we find friends, people with shared interests, and even mates. It determines how we learn about the world and where we get our news; if we get our news and opinions from a focused set of sources that adapt to our interests, that's not good. Indeed, the term *filter bubble* reflects just how influential the web is in shaping our thoughts and opinions.

Along with myriad opportunities and benefits, the web has brought problems and risks because it enables action at a distance—we are visible and vulnerable to far-away people that we have never met.

The web raises unresolved jurisdictional issues. For instance, in the US, many states charge sales tax on purchases within their borders, but online stores often do not collect sales tax from purchasers, on the theory that if they have no physical presence in a state, they are not required to act as an agent of that state's tax authorities. Purchasers are supposed to report out-of-state purchases and pay tax on them, but no one does.

Libel is another area where jurisdiction is uncertain. In some countries it is possible to bring a libel suit merely because a web site (hosted elsewhere) is visible in that country even though the person allegedly doing the libel has never been in the country where the suit is filed.

Some activities are legal in one country but not in another; pornography, online gambling, and criticizing the government are common examples. How does a government enforce its rules on its citizens when they are using the Internet and the web for activities that are illegal within its borders? Some countries provide only a limited number of Internet pathways into and out of the country and can use those to block, filter or slow down traffic that they do not approve of; the Great Firewall of China is the best known example, but certainly not the only one.

Requiring people to identify themselves on the Internet is another approach. It sounds like a good way to prevent anonymous abuse and harassment but it also has a chilling effect on activism and dissent. How do we limit traffic from anonymous trolls and bots while providing anonymity when appropriate?

Attempts by companies like Facebook and Google to force their users to use real names have met with strong resistance, for good reasons. Although there are many drawbacks to online anonymity—hate speech, bullying, and trolling are compelling examples—it's also important for people to be able to express themselves freely without fear of reprisal. We have not found the right balance yet, if indeed one even exists.

There will always be tension between the legitimate interests of individuals, governments (whether they have the support of their citizens or not), and corporations (whose interests often transcend national boundaries). Of course criminals don't worry much about jurisdiction or the legitimate interests of other parties. The Internet makes all of these concerns more pressing.

Part IV

Data

Data is the fourth part of the book. The previous edition was divided into three parts, and data was lumped in with communications. In the past few years, however, data has become so important that it warrants a separate section.

The word "data" often comes qualified—big data, data mining, data science, and a new job title, data scientist. There are books, tutorials, online courses, even university degrees in these topics. Let's take a moment to explain them informally.

Big data means only that we are dealing with a lot of data, which is certainly true enough. Estimates of how much data there is in the world grow ever larger. It used to be that estimates could be conveniently expressed in exabytes (10^{18}) but those days have passed, and now we need zettabytes (10^{21}). It seems safe to predict that yottabytes (10^{24}) lie in the near future. Yotta is the largest prefix in the International System of Units (SI). Eventually, yotta will not be big enough either, and we will have to add some more prefixes, something like "On Beyond Yotta," inspired by Dr. Seuss's *On Beyond Zebra!*.

Data mining is the process of looking for potentially valuable information and insights that can be extracted from all that big data. *Data science* is an interdisciplinary field that applies statistics, machine learning, and other techniques to try to understand the data, extract meaning from it, and make predictions based on it. Naturally a data scientist is a person who does that, and probably hopes to be handsomely paid for working in such a trendy and important field.

Where does all this data come from, what can we do with it, and how can we opt out if we don't want to contribute data about ourselves?

In Chapter 11 we will discuss the myriad sources of data: how our online actions, and our offline actions as well, contribute to what is sometimes called our "data exhaust," the enormous amount of information about us that is produced as we move through the world.

Chapter 12, on artificial intelligence and machine learning, is a look at one aspect of what is done with all those piles of data. Some of it is used for our benefit—image recognition and computer vision, voice recognition and speech processing, language translation, and other useful applications are all made possible because there is so much data to learn from. But the downside is that much can be learned about us as well, often personal information that we don't want anyone to know, or at least to be able to take advantage of.

Extensive use of machine learning raises serious concerns about inferences from data that might support racism, discrimination and other ethical problems. It's tempting to think that machine learning is an objective guide, but in many cases its conclusions merely cloak implicit biases with a veneer of authority.

Chapter 13 discusses defenses: what we can do to limit the data we unwittingly provide and reduce the use that is made of it. It's not possible to be completely invisible or completely safe, but you can significantly improve your personal privacy and security.

11

Data and Information

"When you look at the Internet, the Internet looks back at you."

with apologies to Friedrich Nietzsche, *Beyond Good and Evil*, 1886.

Almost everything you do with your computer, your phone or your credit card generates data about you that is carefully collected, analyzed, saved forever, and often sold to organizations that you know nothing of.

Think about a typical interaction. You use your computer or your phone to search for something to buy, a place to visit or a topic to learn more about. Search engines record what you searched for and when, where you were, and what results you clicked on, and if possible they associate it with you specifically. Advertisers use that information to send you targeted messages about their offerings.

All of us search, shop, and entertain ourselves with online movies and TV shows. We communicate with friends and family by mail and text or even occasionally a voice call. We use Facebook or Instagram to stay in touch with friends and acquaintances, LinkedIn to maintain potential job connections, and perhaps dating sites to find romance. We read Reddit, Twitter and online news to keep up with the world around us. We manage our money and pay our bills online. We move around constantly with a phone that knows exactly where we are at all times. Our cars know where we are and relay that information to others. Of course ubiquitous cameras also know where our cars are. Home systems like networked thermostats, security systems and smart appliances watch our every move, and know whether we're at home, and what we're doing when we are.

Every bit of this stream of personal data is collected. A 2018 prediction by Cisco, the dominant manufacturer of network hardware, says that annual global Internet traffic will exceed 3 zettabytes in 2021. The prefix *zetta* is 10^{21}, which is a lot of bytes by any measure. Where does all this data come from and what's being done with it? The answers are sobering, since most of that data is not for us but about us. The more data there is, the more that strangers can learn about us and the more that our privacy and security are decreased.

I'll start with web search, since a significant amount of data collection begins with search engines. That leads to a discussion of tracking—what sites you have visited and what you did while visiting them. Next, I'll talk about the personal information people willingly give away or trade for entertainment or a convenient service. Where is all that data kept? That leads to databases—collections of data that have been gathered by all kinds of parties—and to data aggregation and mining, because much of the value of data comes when it is combined with other data to give new insights. That's also where major privacy problems arise, since combining data about us from multiple sources makes it all too easy to learn things that are no one else's business. Finally, I'll discuss cloud computing, where we hand over everything to companies that provide storage and processing on their servers instead of on our own computers.

11.1 Search

Web search began in 1995 when the web was tiny by today's standards. The number of web pages and the number of queries both rose rapidly over the next few years. The original paper on Google, "The anatomy of a large-scale hypertextual web search engine," by Sergey Brin and Larry Page, was published in early 1998. It said that AltaVista, one of the most successful of the early search engines, was handling 20 million queries per day late in 1997, and accurately predicted a web of a billion pages and hundreds of millions of queries a day by 2000. One estimate is that in 2017 there were 5 billion queries a day.

Search is big business, going from nothing to a major industry in less than 20 years. For example, Google was founded in 1998, went public in 2004, and had a market capitalization of a trillion dollars in fall 2020, way behind Apple (over $2 trillion) but far ahead of long-established businesses like Exxon Mobil and AT&T, both under $200 billion. Google is highly profitable but there's plenty of competition, so who knows what might happen. (A disclosure is in order here: I am a part-time Google employee and I have many friends at the company. Naturally, nothing in this book should be taken as Google's position on anything.)

How does a search engine work? From a user's perspective, a query is typed into a form on a web page and sent to a server, which almost instantly returns a list of links and text snippets. From the server side, it's more complicated. The server creates a list of pages that contain the word or words of the query, sorts them into order of relevance, wraps snippets of the pages in HTML, and sends them back.

The web is far too big for each user query to trigger a new search over the web, however, so a major part of a search engine's task is to be prepared for queries by already having page information stored and organized at the server. This is done by a *crawl* of the web, which scans pages and stores relevant contents in a database so that subsequent queries can be answered quickly. Crawling is a giant-scale example of caching: search results are based on a pre-computed index of cached page information, not a real-time search of Internet pages.

Figure 11.1 shows approximately what the structure is, including the insertion of advertising into the result page.

The problem is scale. There are billions of users and many billions of web pages. Google used to report how many pages it was crawling to build its index, but stopped

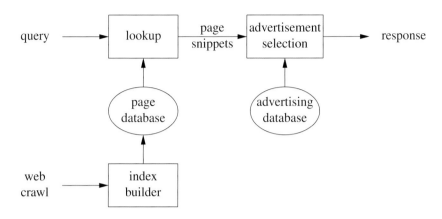

Figure 11.1: Organization of a search engine.

doing that sometime after the number of pages went past 10 billion. If a typical web page is 100 KB, storing a hundred billion pages requires 10 petabytes of disk space. Although some of the web is static pages that don't change for months or even years, a significant fraction changes rapidly (news sites, blogs, Twitter feeds), so the crawl has to be ongoing and highly efficient; there's no chance to rest, lest the indexed information become dated. Search engines handle billions of queries a day; each of those requires scanning a database, finding the relevant pages, and sorting them into the right order. It also includes choosing advertisements to accompany the search results and, in the background, logging everything so there's data to improve search quality, keep ahead of competitors, and sell more advertising.

From our standpoint, a search engine is a great example of algorithms in action, though the volume of traffic means that no simple search or sort algorithm is going to be fast enough.

One family of algorithms deals with the crawl: deciding what page to look at next, extracting the indexable information from it (words, links, images, and so on), and delivering those to the index builder. URLs are extracted, duplicates and irrelevancies are eliminated, and the survivors are added to a list to be examined in their turn. Complicating this is the fact that a crawler can't visit a site too often, because that could add a significant load and thus become a nuisance; the crawler might even be denied access. Since the rate of change of pages varies widely, there is also a payoff from algorithms that can accurately assess how frequently pages change, and visit the rapid changers more often than the slow changers.

Index-building is the next component. This takes pages from the crawler, extracts the relevant parts of each page, and indexes each part along with its URL and location in the page. The specifics of this process depend on the content to be indexed: text, images, spreadsheets, PDFs, videos and so on all require their own processing. In effect, indexing creates a list of pages and locations for each word or other indexable item that has occurred in some web page, stored in a form that makes it possible to quickly retrieve the list of pages for any specific item.

The final task is formulating a response to the query. The basic idea is to take all the words of the query, use the indexing lists to quickly find URLs that match, then (also quickly) select the URLs that match best. The details of this process are the crown jewels for search engine operators, and you won't find out much about specific techniques by looking on the web. Again, scale dominates: any given word might appear in millions of pages, a pair of words might still be in a million pages, and those potential responses have to be rapidly winnowed down to the best ten or so. The better that the search engine brings accurate hits to the top and the more rapidly it responds, the more it will be used in preference to its competitors.

The first search engines merely displayed a list of pages containing the search terms, but as the web grew, search results were a jumble of largely irrelevant pages. Google's original *PageRank* algorithm assigned a quality measure to each page. PageRank gave more weight to pages that other pages link to or that are linked to by pages that themselves are highly ranked, on the theory that those are the most likely to be relevant to the query. As Brin and Page say, "Intuitively, pages that are well cited from many places around the web are worth looking at." Naturally there's far more than this to producing high-quality search results, and search engine companies are continuously trying to improve their results in comparison to their competitors.

It takes enormous computing resources to provide a search service at full scale—millions of processors, terabytes of primary memory, petabytes of secondary storage, gigabits per second of bandwidth, gigawatts of electrical power, and of course lots of people. These all have to be paid for somehow, usually by advertising revenue.

At the most basic level, advertisers pay to display an advertisement on a web page, with the price determined by some measure of how many people, and what kind of people, view the page. The price might be in terms of page views ("impressions," which count only that the ad appeared on the page), clicks (the viewer clicked on the ad), or "conversions," where the viewer eventually bought something. Viewers who might be interested in whatever is being advertised are clearly valuable, so in the most common model, the search engine company runs a real-time auction for search terms. Advertisers bid for the right to display advertisements beside the results for specific search terms, and the company that serves the ad makes money when a viewer clicks on it.

Google Ads (formerly AdWords) makes it easy to experiment with a proposed advertising campaign. For instance, their estimation tool (Figure 11.2) says that the likely cost for the search terms "kernighan" and some related ones like "unix" and "c programming" will be 5 cents per click, that is, every time someone searches for one of these terms and then clicks on my advertisement, I will pay Google 5 cents. Google also estimates that for my selection of search terms there will be 194 clicks per day with a daily budget of ten dollars (averaged over a month), though of course neither of us knows how many people would click on my advertisement and thus cost me money. I've never done the experiment to see what would happen.

Could advertisers pay to bias the search results in favor of themselves? This was a concern for Brin and Page, who wrote in the same paper, "We expect that advertising funded search engines will be inherently biased towards the advertisers and away from the needs of the consumers." Google derives most of its revenue from advertising. Though it maintains a separation between search results and advertisements, as do the other major search engines, numerous legal challenges have claimed unfairness or bias towards Google's own products. Google's response is that search results

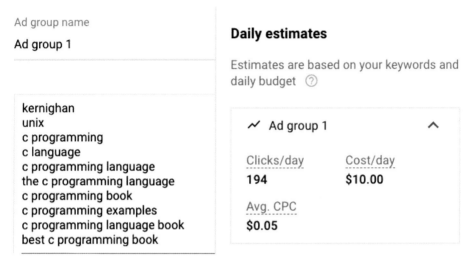

Figure 11.2: Google Ads estimates for "kernighan" and related terms.

are not biased against competitors but are based entirely on algorithms that reflect what people find most useful.

Another possible form of bias occurs when nominally neutral advertising results are subtly biased towards particular groups based on possible profiling for race, religion, or ethnicity. For instance, some names are predictive of racial or ethnic background, so advertisers could aim towards or away from such groups when the names are searched for.

In the US, some kinds of advertisements are illegal if they indicate a preference based on race, religion, or gender. Facebook, which also derives almost all of its revenue from advertising, provides advertisers with tools to target their advertisements using a large set of criteria, most of which are straightforward—income, education— but some are explicitly illegal and others are proxies for potentially discriminatory targeting. In 2019, Facebook settled a suit that alleged that its advertising platform permitted advertisements that enabled discrimination.

Is it possible to search the web without being tracked in such detail? The Duck-DuckGo search engine promises that it does not maintain your personal search history and does not deliver personalized advertisements. It does some of its own searching and aggregates results from a large number of search engines and other sources. It makes money from advertising, but that can be removed by Adblock and the like. DuckDuckGo also offers several useful guides for more private and secure browsing and phone use.

11.2 Tracking

The discussion above is phrased in terms of search, but of course the same kinds of considerations apply for any kind of advertising: the more accurately it can be targeted, the more likely it is to cause the viewer to respond favorably and thus the more an advertiser is willing to pay. Tracking you online—what you search for, what sites you visit, and what you do while visiting them—can reveal a remarkable amount about who you are and what you do. For the most part, the goal of today's tracking is to sell to you more effectively, but it's not hard to imagine other uses of such detailed information. In any case, the focus in this section is primarily on tracking mechanisms: cookies, web bugs, JavaScript, and browser fingerprinting.

It's inevitable that information about us is collected as we use the Internet; it's hard to do anything without leaving tracks. The same is true when we use other systems, especially cell phones, which know our physical location whenever they are turned on. A GPS-enabled phone, which includes all smartphones, generally knows where you are to within about 10 meters when you're outside, and can report your position at any time. Some digital cameras include GPS as well, which lets them encode the geographic location in each picture they take; this is called *geo-tagging*. Cameras can use Wi-Fi or Bluetooth for uploading pictures; there's no reason your camera couldn't be used for tracking as well.

When tracks like these are collected from multiple sources, they paint a detailed picture of our activities, interests, finances, associates, and many other aspects of our lives. In the most benign setting, that information is used to help advertisers target us more accurately, so we will see advertisements that we are likely to respond to. But the tracking need not stop there and its results can be used for much less innocent purposes, including discrimination, financial loss, identity theft, surveillance, and even physical harm.

In 2019 and 2020, the *New York Times* published an extended series of articles on privacy and tracking. One of the most revealing and disturbing was a study of a database of 50 billion cell phone location records from the phones of 12 million people in several large American cities. The data was provided by anonymous sources, most likely someone working at a data broker. To quote the *Times*,

> The companies that collect all this information on your movements justify their business on the basis of three claims: People consent to be tracked, the data is anonymous and the data is secure.

> None of those claims hold up.

The *Times* was able to identify a significant number of individuals by correlating with events, home and work addresses, and the like. The *Times* was working with 50 billion records, but says that location data companies collect orders of magnitude more information than that *every day*, including a great deal of demographic data, so correlations and identifications are easier. In theory, there is no personally identifiable information in this "anonymous" data, but in practice, it is easy to make connections that identify individuals, especially when data sources are combined. The article is seriously worrying, as is the whole series.

How is information collected? Some information is sent automatically with every request from your browser, including your IP address, the page that you were viewing (the "referrer"), the type and version of your browser (the "user agent"), the operating system, and your language preference. You have only limited control over this. Figure 11.3 shows some of the information that is sent, edited for space.

In addition, if there are cookies from the server's domain, those are sent as well. As discussed in the last chapter, cookies only go back to the domain from which they originated, so how could one site use cookies to track visits to other sites?

```
HTTP_ACCEPT text/html,application/xhtml+xml,application/xml;
        q=0.9,image/webp,*/*;q=0.8
HTTP_ACCEPT_ENCODING gzip, deflate
HTTP_ACCEPT_LANGUAGE en-US,en;q=0.5
HTTP_CONNECTION keep-alive
HTTP_DNT 1
HTTP_HOST [...].princeton.edu
HTTP_REFERER http://[...].princeton.edu/env.html
HTTP_USER_AGENT Mozilla/5.0 (Windows NT 10.0;
        rv:68.0) Gecko/20100101 Firefox/68.0
QUERY_STRING [...]
REMOTE_ADDR 128.112.139.195
TZ America/New_York
```

Figure 11.3: Some of the information sent by a browser.

The answer is implicit in how links work. Web pages contain links to other pages—that's the essence of hyperlinking—and we're familiar with links that must be explicitly clicked before they're followed. But image and script links do not need to be clicked; they are loaded automatically from their sources as the page is being loaded. If a web page contains a reference to an image, that image is loaded from whatever domain is specified. Normally the image URL will encode the identity of the page making the request so that when my browser fetches the image, the domain that provides the image knows which page I have accessed and it too can deposit cookies on my computer or phone and retrieve cookies from previous visits. The same is true of JavaScript scripts.

This is the heart of tracking, so let's make it more specific. As an experiment I turned off all my defenses and used Safari to visit `toyota.com`. The initial visit downloaded cookies from over 25 different sites, along with 45 images from a wide variety of sites and over 50 JavaScript programs, totaling more than 10 megabytes.

The page continued to make network requests for as long as I remained on it, and in fact was doing so much background computing that Safari warned me about it (Figure 11.4).

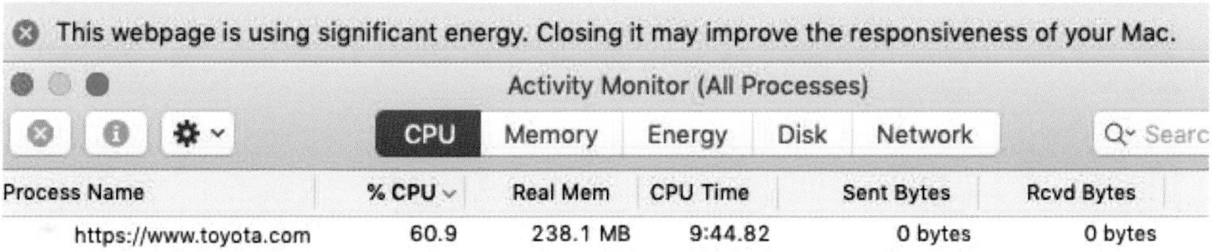

Figure 11.4: The web page that keeps on computing.

This explains why when I ask my students to count cookies, they report that they have thousands. It also explains why such pages might load slowly. (You can do your own experiments; this information comes from places like the browser's history and privacy settings.) I didn't try the experiment on a phone, since it would make a measurable dent in my modest data plan.

With my normal defenses enabled—Ghostery, Adblock Plus, uBlock Origin, NoScript, no cookies, no local data storage—I get no cookies or scripts at all.

A significant number of the images in the page were like the one highlighted in Figure 11.5. The Toyota web page includes a link to Facebook that fetches an image. The image is 1 pixel wide by 1 pixel high and is transparent, so it's completely invisible.

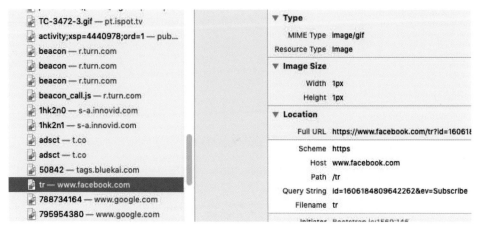

Figure 11.5: A single pixel image for tracking.

Single-pixel images like this are often called *web bugs* or *web beacons*; their sole purpose is tracking. When my browser requests the image from Facebook, Facebook knows that I am looking at a particular Toyota page and (if I allow it) can store a cookie. As I visit other sites, each tracking company can build up a picture of what I'm looking at. If it's mostly about cars, they can tell prospective advertisers and I'll start to see advertisements for car dealers, loans, and automotive accessories. If it's about accidents and pain relief, I might see more from repair services, lawyers and therapists.

Companies like Google, Facebook, and myriad others collect information about sites that people have visited and then use that to sell advertising space to customers like Toyota, who can in turn use it for targeted advertising and (perhaps) correlation with other information about me beyond my IP address. As I visit more web pages, tracking companies create ever more detailed databases of my presumed characteristics and interests, and may eventually deduce that I'm male, married, over 60, own a couple of cars, live in central New Jersey and work at Princeton University. The more they know about me, the more accurately their customers can target their advertisements. Of course targeting *per se* is not the same as identification, but at some point it should be easy to identify me personally, though most such companies say that they don't do that. If I do give my name or email address on some web page, however, there's no guarantee that it won't be passed around.

In 2016, the *Washington Post* published a series of stories on privacy. One article was entitled "98 personal data points that Facebook uses to target ads to you." The 98 data points include obvious things like location, age, gender, language, education level, income and net worth, but also touchier ones like "ethnic affinity" that could be used for illegal discrimination.

Internet advertising is a sophisticated marketplace. When you request a web page, the web page publisher notifies an advertising exchange like Google Ad

Exchange or AppNexus that space on that page is available, and provides information about the likely viewer, perhaps a 25–40-year-old single woman in San Francisco who likes technology and good restaurants. Advertisers bid for the ad space and the winner's advertisement is inserted into the page, all in a few hundred milliseconds.

If being tracked does not appeal to you, it can be significantly reduced, though not without some cost. Browsers allow you to reject cookies entirely or to disable third-party cookies. You can remove cookies explicitly any time and they can be removed automatically when the browser is closed. Major tracking companies provide an opt-out mechanism: if they encounter a specific cookie on your computer, they will not track your interactions for targeted advertising, though it's highly likely that they will still track you on their own sites.

There is a quasi-official Do Not Track mechanism that promises more than it delivers. Browsers have a checkbox, usually within the privacy and security menu, called "Do Not Track," which if set causes an extra HTTP header to be sent with requests. (Figure 11.3 includes an example.) A web site that honors DNT will not pass on information about you to other sites, though it's free to retain information for its own use. In any case, respecting the visitor's wishes is entirely voluntary and most sites ignore the preference. For instance, Netflix says "At this time, we do not respond to Web browser 'do not track' signals."

Private browsing or *incognito mode* is a client-side mechanism that tells the browser to clear history, cookies, and other browsing data when the browser session terminates. That prevents other users of your computer from knowing what you have been doing (which is why it's informally known as "porn mode"), but it has no effect whatsoever on what is remembered at the sites you have visited, which with high probability can recognize you again anyway. In spite of that, some sites refuse to provide content if they detect that you are using incognito mode.

Defense mechanisms are not standardized across browsers or even across different versions of the same browser, and the defaults are usually set so you are defenseless.

Unfortunately, many sites don't work without cookies, though most work fine without third-party cookies, so you should always turn those off. Some uses of cookies are legitimate—a web site needs to know that you have logged in already or wants to keep track of your shopping cart—but often cookies are used for tracking. This irritates me enough that I prefer not to patronize such places.

JavaScript is a major tracking tool. A browser will run any JavaScript code found in the original HTML file or loaded from a URL with `src="name.js"` within a `<script>` tag. This is heavily used for "analytics" that watch how particular pages are viewed. For example, when I visit Slashdot.org, a technology news site, my browser downloads 150 KB for the page itself, but it also downloads another 115 KB of JavaScript analytics scripts from three other sites, including this one from Google:

```
<script>
  src="https://google-analytics.com/ga.js">
</script>
```

(None of these analytic scripts are actually downloaded when I personally visit Slashdot, since I use extensions like Adblock and Ghostery to block them.)

JavaScript code can set and retrieve cookies from the site that the JavaScript itself came from, and can access other information like the browser's history of pages visited. It can also monitor the position of your mouse continuously and report that back to a server, which can infer what parts of the web page might be interesting or not; it can also monitor places where you clicked or highlighted even if they weren't sensitive areas like links.

Figure 11.6 shows a couple of lines of JavaScript that will display the position of the mouse as you move it. A few more lines would send that same information to the supplier of whatever web page you were viewing, along with other events like where you typed, clicked or dragged. The site clickclickclick.click is a much more polished and highly entertaining version of the same idea.

```
<html>
<script>
function move(event) {
  document.getElementById("body").innerHTML =
    "position: " + event.clientX + " " + event.clientY;
}
</script>
<body>
  <div id="body" style="width:100%; height: 500px;"
       onmousemove="move(event)">
  </div>
</body>
</html>
```

Figure 11.6: JavaScript to display coordinates as the mouse moves.

Browser fingerprinting uses individual characteristics of your browser to identify you, often uniquely, without cookies. The combination of operating system, browser, version, language preference, installed fonts and plug-ins provides a lot of distinctive information. Using new facilities in HTML5, it's possible to see how an individual browser renders specific character sequences, using a technique called *canvas fingerprinting*. A handful of these identifying signals is enough to distinguish and recognize individual users regardless of their cookie settings. Naturally advertisers and other organizations would love to have pinpoint identification of individuals whether they disable cookies or not.

The Electronic Frontier Foundation (EFF) offers an instructive service called Panopticlick, after Jeremy Bentham's "Panopticon," a prison designed so that inmates could be continuously monitored without their knowing when they were being watched. Visiting panopticlick.eff.org will tell you approximately how unique you are among other recent visitors. Even with good defenses, you are likely to be uniquely identifiable or at least close to it. With high probability, you will be recognized when they see you the next time.

Blacklight, which you can find at themarkup.org/blacklight, simulates a defenseless browser and reports back on trackers (including those that are trying to evade ad-blockers), third-party cookies, mouse and keyboard monitoring, and other devious practices. It's sometimes scary to see how much tracking goes on, and it can be entertaining to try to find the worst offenders. For example, the cooking site epicurious.com loaded 136 third-party cookies and 44 ad trackers, while monitoring keystrokes and mouse clicks, and reporting the visit to Facebook and Google.

Tracking mechanisms are not restricted to browsers, but can be used by mail readers and other systems. If your mail reader interprets HTML, then it's going to "display" those single-pixel images that let someone track you. Apple TV, Chromecast, Roku, TiVo, and Amazon's Fire TV stick all know what you're watching. So-called "smart TVs" know that too, but may also send your voice back to the manufacturer, and even images from their cameras. Speech-enabled devices like Amazon Echo send what you say off for analysis.

As we saw earlier, every IP packet travels through 15 to 20 gateways on its way from your computer to its destination, and the same is true of packets coming back. Each gateway along that path can inspect each packet to see what it contains and even modify the packet in some way. This is called *deep packet inspection* because it looks not just at the headers of information but the actual contents. Usually this happens at your ISP, since that's the place where you are most easily identifiable. It's not limited to web browsing, but includes all traffic between you and the Internet.

Deep packet inspection can be used for valid purposes, like weeding out malware, but it can also be used to better aim targeted advertising, or to monitor or interfere with traffic into and out of a country, like the Great Firewall of China, or the taps that the NSA placed on traffic in the US.

The only defense against deep packet inspection is end-to-end encryption with HTTPS, which protects the contents from inspection and modification as it travels, though it does not hide metadata like source and destination.

The rules governing what personally identifiable information can be collected and how it can be used vary from country to country. As an over-simplification, in the US, anything goes—any company or organization can collect and distribute information about you, without notice and without offering you a chance to opt out.

In the European Union (again, over-simplified), privacy is taken more seriously: companies cannot legally collect or use data about an individual without the individual's explicit permission. The relevant part of the General Data Protection Regulation, or GDPR, which went into effect in mid-2018, says that individual personal data cannot be processed unless consent has explicitly been given. Even an online form with opt-out as default is not deemed sufficient consent. People also have the right to access their personal information and to see how it is being used. Consent can be withdrawn at any time.

The US and the EU set up an agreement in 2016 to govern how data could be moved between the two regions while protecting the privacy rights of EU citizens. In July 2020, however, the top EU court ruled that the agreement did not comply with EU privacy rights, leaving the current situation unclear.

The California Consumer Privacy Act (CCPA) went into effect at the beginning of 2020, with goals and properties similar to the GDPR. It includes an explicit "do not sell my data" option. Although it applies only to residents of California, one might hope that it will have wider effects in the US. California has over 10 percent of the US population, and is often ahead of the curve in social issues.

It's still too early, however, to know whether the GDPR and the CCPA are working well.

11.3 Social Networks

Tracking the web sites we visit is not the only way that information about us can be gathered. Indeed, users of social networks *voluntarily* give up a remarkable amount of personal privacy in return for entertainment and keeping in touch with other people.

Years ago, I came across a web post that went something like this: "In a job interview, they asked me questions about things that weren't mentioned in my resume. They had been looking at my Facebook page, which is outrageous because Facebook is about my private life and has nothing to do with them." This is touchingly naive and innocent, but one suspects that at least some Facebook users would feel similarly violated even though it's well known that employers and college admissions offices routinely use search engines, social networking and similar sources to learn more about applicants. In the US, it's illegal to ask a job applicant about age, ethnicity, religion, sexual orientation, marital status, and a variety of other personal information, but those can be readily, and quietly, determined by searching social networks.

Search engines and social networks provide useful services, and they are free; what's not to like? But they have to make money somehow, and you should remember that if you are not paying for a product, you *are* the product. The business model of social networking sites is to collect a great deal of information about their users and sell it to advertisers. Accordingly, almost by definition, they are going to have privacy problems.

In the short time they have existed, social networks have grown dramatically in size and influence. Facebook was founded in 2004 and in 2020 claimed over 2.5 billion active users each month, about one third of the world's population. (Facebook also owns Instagram and WhatsApp, and information is shared among the operations.) Its annual revenues of $70 billion come almost entirely from advertising. That growth rate doesn't allow much time for careful consideration of policies, nor does it allow for leisurely development of robust computer programs. Every social networking site has had problems with divulging private information through ill-considered features, user confusion about privacy settings (which change frequently), software errors, and data disclosures that are inherent in the overall system.

As the largest and most successful social network, Facebook's issues have been most visible. Some have come about because Facebook provides an API for third parties to write applications that run in the context of a Facebook user, and those can reveal private information contrary to the company's privacy policies. Again, none of this is unique to Facebook.

Geolocation services display users' locations on cell phones, which makes it easy to meet friends in person or to play location-based games. Targeted advertising can be especially effective when a potential customer's physical location is known; you're more likely to respond to an offer from a restaurant when you're standing outside its door than reading about it in a newspaper. On the other hand, it's a bit creepy when you realize that your phone is being used to track you around the interior of a store. Nevertheless, stores are beginning to use *in-store beacons*. If you opt in to the system, usually by downloading a specific app and implicitly agreeing to be tracked, the beacons, which use Bluetooth to communicate with the app on your phone, monitor your location within the store and will offer you deals if it looks like you might be tempted by something specific. To quote one company that makes beacon systems, "beacons are ushering in the indoor mobile marketing revolution."

Location privacy—the right to have your location remain private—is compromised by systems like credit cards, toll payment systems on highways and public transport, and of course cell phones. It's harder and harder to avoid leaving a trail of every place you've ever been. Cell phone apps are the worst offenders, often requesting access to essentially everything that the phone knows about you, including call data, physical location, and so on. Does a flashlight app really need my location, contacts and call log?

Intelligence agencies have known for a long time that they can learn a great deal by analyzing who communicates with whom, without knowing what the parties said. It's for this reason that the NSA has been collecting metadata on all telephone calls in the US—phone numbers, time and duration. The initial collection was authorized as one of the hasty responses to the terrorist attacks on September 11, 2001, but the extent of data collection was not appreciated until the release of Snowden's documents in 2013. Even if one accepts the claim that "it's just metadata, not the conversations," metadata can be exceptionally revealing. In testimony before a Senate Judiciary Committee hearing in October 2013, Ed Felten of Princeton University explained how metadata can make a private story completely public:

> Although this metadata might, on first impression, seem to be little more than 'information concerning the numbers dialed,' analysis of telephony metadata often reveals information that could traditionally only be obtained by examining the contents of communications. That is, metadata is often a proxy for content.

> In the simplest example, certain telephone numbers are used for a single purpose, such that any contact reveals basic and often sensitive information about the caller. Examples include support hotlines for victims of domestic violence and rape. Similarly, numerous hotlines exist for people considering suicide, including specific services for first responders, veterans, and gay and lesbian teenagers. Hotlines exist for sufferers of various forms of addiction, such as alcohol, drugs, and gambling.

> Similarly, inspectors general at practically every federal agency—including the NSA—have hotlines through which misconduct, waste, and fraud can be reported, while numerous state tax agencies have dedicated hotlines for reporting tax fraud. Hotlines have also been established to report hate crimes, arson, illegal firearms and child abuse. In all these cases, the metadata alone conveys a great deal about the content of the call, even without any further information.

> The phone records indicating that someone called a sexual assault hotline or a tax fraud reporting hotline will of course not reveal the exact words that were spoken during those calls, but phone records indicating a 30-minute call to one of these numbers will still reveal information that virtually everyone would consider extremely private.

The same is true of the explicit and implicit connections in social networks. It's much easier to make linkages among people when the people provide the links explicitly. For example, Facebook "Likes" can be used to accurately predict characteristics like gender, ethnic background, sexual orientation and political leaning. This indicates the kind of inferences that can be made from information freely given away by users of social networks.

Facebook "Like" buttons and similar ones from Twitter, LinkedIn, YouTube and other networks make tracking and association much easier. Clicking on a social icon on a page reveals that you are looking at the page—it's in effect an advertising image, though visible instead of hidden, and it gives the supplier a chance to send a cookie.

Personal information leaks from social networking and other sites even for non-users. For instance, when I get an electronic invitation (an "e-vite") to a party from a well-meaning friend, the company that runs the invitation service now has a guaranteed email address for me, even though I do not respond to the invitation or in any way give permission to use my address.

If a friend tags me in a picture that is posted on Facebook, my privacy has been compromised without my consent. Facebook offers face recognition so that friends can tag each other more easily, and the default allows tagging without the taggee's permission. It appears that my only control is that I can opt out of having Facebook suggest that I be tagged but not out of tagging itself; according to Facebook:

> When you turn your face recognition setting on, we'll use face recognition technology to analyze photos and videos we think you're in, such as your profile picture and photos you've already been tagged in, to create a template for you. We use your template to recognize you in other photos, videos and other places where the camera is used (like live video) on Facebook.
>
> When you turn this setting off: [...] We won't use face recognition to suggest that people tag you in photos. This means that you'll still be able to be tagged in photos, but we won't suggest tags based on a face recognition template.

I don't use Facebook at all and thus was surprised to find that I "have" a Facebook page, apparently generated automatically from Wikipedia. This irritates me but there's little that I can do about it, other than hoping that people don't think I endorse it.

It's easy for any system with a large user population to create a "social graph" of interactions among its direct users and to include those who are brought in indirectly without their consent or even knowledge. In all of these cases, there's no way for an individual to avoid the problem ahead of time and it's hard to remove the information once it has been created.

Think hard about what you tell the world about yourself. Before sending mail or posting or tweeting, pause a moment and ask whether you would be comfortable if your words or pictures appeared on the front page of the *New York Times* or as the lead story on a TV news program. Your mail, texts and tweets are likely to be stored forever and could potentially reappear in some embarrassing context years later.

11.4 Data Mining and Aggregation

The Internet and the web have caused a revolution in how people collect, store and present information. Search engines and databases are of enormous value to everyone, to the point where it's hard to remember how we managed before the Internet. Massive amounts of data ("big data") provide the raw material for speech recognition, language translation, credit card fraud detection, recommendation systems, real-time traffic information, and many other invaluable services.

But there are also major downsides to the proliferation of online data, especially for information that might reveal more about ourselves than is comfortable.

Some information is clearly public and some collections of information are meant to be searched and indexed. If I create a web page for this book, it's definitely to my advantage to have it easily found by search engines.

What about public records? Legally, certain kinds of information are available to any member of the public upon application. In the United States, this includes court proceedings, mortgage documents, house prices, local property taxes, birth and death records, marriage licenses, voter rolls, political contributions, and the like. (Note that birth records reveal date of birth and are likely to expose the "mother's maiden name," which are often used as part of verifying one's identity.)

In olden times it required a trip to a local government office to get this information, so although the records were technically "public," they were not accessible without some effort; the person seeking the data had to show up in person, perhaps identify himself or herself, and maybe pay for each physical copy. Today the data is often available online and I can examine public records anonymously from the comfort of my home. I might even be able to run a business by collecting them in volume and combining them with other information. The popular site `zillow.com` combines maps, real estate advertisements, and public data on properties and transactions to display house prices on a map. This is a valuable service if one is looking to buy or sell, but otherwise might be thought intrusive. Similar sites add information on current and past residents, their voter registration information, and as a teaser, hint at potential criminal records. The Federal Election Commission database of election contributions at `fec.gov` shows which candidates have been supported by friends and notables, and can reveal information like home addresses. This strikes an uneasy balance that leans more towards the public's right to know than to the individual's right to privacy.

Questions about what information should be so accessible are hard to answer. Political contributions should be public but home addresses of contributors should probably be redacted. Personal identifiers like US Social Security numbers should never be put on the web since they are too readily used for identity theft. Arrest records and photographs are sometimes public and there are sites that display that information; their business model is to charge individuals for removing the pictures! Existing laws don't always prevent the release of such information and the horse has left the barn in many cases—once on the web, information is likely to always be on the web.

Concerns about freely available information become more serious when data is combined from multiple sources that might seem independent of each other. For instance, companies that provide web services record a great deal of information about their users. Search engines log all queries, along with the IP address from which the query came and cookies from a previous visit.

In August 2006, with the best of intentions, AOL released a large sample of search logs for research. The logs of 20 million queries by 650,000 users over three months had been anonymized, so that in theory anything that could have identified individual users was completely removed. Good intentions notwithstanding, it rapidly became clear that in practice the logs were not nearly as anonymized as AOL had thought. Each user had been assigned a random but unique identification number, and this made it possible to find sequences of queries made by the same person, which in turn made it possible to identify at least a few individuals uniquely. People had searched for their own names, addresses, Social Security numbers, and other personal information. Correlation of searches revealed more than AOL had believed and certainly much more than the original users would have wanted. AOL quickly removed the query logs from its web site, but of course that was far too late; the data had spread all over the world.

Query logs contain valuable information for running a business and improving the service, but clearly they also contain potentially sensitive personal information. How long should search engines retain such information? There are conflicting external pressures here: a short period for privacy reasons versus a long period for law enforcement purposes. How much should they process the data internally to make it more anonymous? Some companies claim to remove part of the IP address for each query, typically the rightmost byte, but that may not be enough to de-identify users. What access should government agencies have to this information? How much would be discoverable in a civil action? The answers to these questions are far from clear. Some of the queries in the AOL logs were scary—questions about how to kill one's spouse, for example—so it might be desirable to make logs available to law enforcement in restricted situations, but where to draw the lines is most uncertain. Meanwhile, there are a handful of search engines that say they do not keep query logs; DuckDuckGo is the most widely used.

The AOL story illustrates a general problem: it's hard to truly anonymize data. Attempts to remove identifying information tend to take a narrow view: there's nothing in this particular data that could identify a specific person so it must be harmless. However, in the real world there are other sources of information, and it is often possible to make deductions by combining facts from multiple sources that might have been completely unknown to the original providers and perhaps did not even exist until later.

A famous early example brought this re-identification problem to the surface vividly. In 1997, Latanya Sweeney, at the time a PhD student at MIT, analyzed ostensibly de-identified medical records for 135,000 Massachusetts state employees. The data had been released by the state's insurance commission for research and was even sold to a private company. Among many other things, each record included birth date, gender and current ZIP code. Sweeney found six people born on July 31, 1945; three were male, and only one lived in Cambridge. Combining this information with public voter registration lists, she was able to identify that person as William Weld, the governor of Massachusetts at the time.

These are not isolated incidents. In 2014, the New York City Taxi and Limousine Commission released an anonymized dataset of all 173 million taxi rides in the city in 2013. But this wasn't done well, so it was possible to reverse-engineer the process and thus re-attach the information about which cab matched each trip, based on the cab's medallion number. At this point, an enterprising data science intern discovered that he could find pictures of famous people getting into taxis where the medallion number was visible. This was enough to reconstruct details of about a dozen trips, right down to the amount tipped.

It's tempting to believe that no one can figure out some secret because they don't know enough. These examples illustrate that it's often possible to learn a great deal by combining datasets that were never meant to be examined together. Adversaries may already know more than you think, and even if they don't today, more information will become available over time.

11.5 Cloud Computing

Think back to the model of computation described in Chapter 6. You have a personal computer, or several. You run separate application programs for different tasks, like Word for creating documents, Quicken or Excel for personal finance, and iPhoto for managing your pictures. The programs run on your own computer, though they might contact the Internet for some services. Every so often you can download a new version that fixes bugs, and occasionally you might have to buy an upgrade to get new features.

The essence of this model is that the programs and their data live on your own computers. If you change a file on one computer and then need it on another, you have to transfer it yourself. If you need a file that's stored on a computer at home while you're in your office or away on a trip, you're out of luck. If you need Excel or PowerPoint on both a Windows PC and a Mac, you have to buy the program for each. And your phone isn't part of this at all.

A different model is now the norm: using a browser or a phone to access and manipulate information that is stored on Internet servers. Mail services like Gmail or Outlook are the most widespread example. Mail can be accessed from any computer or phone. It's possible to upload a mail message that was composed locally or to download messages to a local file system, but mostly you just leave the information with whoever provides the service. There's no notion of software update, though from time to time new features appear. Keeping up with friends and looking at their pictures is often done with Facebook; the conversations and pictures are stored at Facebook, not on your own computer. These services are free; the only visible "cost" is that you might see advertisements as you read your mail or check in to see what your friends are doing.

This model is often called *cloud computing*, because of the metaphor that the Internet is a "cloud" (Figure 11.7) with no particular physical location, and information is stored somewhere "in the cloud." Mail and social networks are the most common cloud services, but there are plenty of others, like Dropbox, Twitter, LinkedIn, YouTube, and online calendars. Data is not stored locally but in the cloud, that is, by the service provider—your mail and your calendar are on Google servers, your

Figure 11.7: The cloud.

pictures are on Dropbox or Facebook servers, your resume is at LinkedIn, and so on.

Cloud computing is made possible by the alignment of a number of factors. As personal computers have become more powerful, so have browsers; browsers can now efficiently run large programs with intensive display requirements, even though the programming language is interpreted JavaScript. Bandwidth and latency between client and server are much better now for most people than they were ten years ago, and this makes it possible to send and receive data quickly, even responding to individual keystrokes for suggesting search terms as you type. As a result, a browser can handle most of the user interface operations that would have required a standalone program in the past, while using a server to hold the bulk of the data and do any heavy computing. This organization also works well for phones: there's no need to download an app.

A browser-based system can be nearly as responsive as a desktop system, while providing access to data from anywhere. Consider the cloud-based "office" tools from Google, which provides a word processor, spreadsheet, and presentation program that allow simultaneous access and update by multiple users.

One of the interesting issues is whether these cloud tools will ultimately run well enough to cut the ground out from under desktop versions. As might be imagined, Microsoft is concerned because Office provides a significant fraction of the company's revenue, and because Office runs primarily on Windows, which provides much of the rest of the revenue. Browser-based word processors and spreadsheets do not need anything from Microsoft and thus threaten both core businesses. At the moment, Google Docs and similar systems do not provide all the features of Word, Excel and PowerPoint, but the history of technological advance is replete with examples where a clearly inferior system appeared, picked up new users for whom it was good enough, and gradually ate away at the incumbent. Microsoft is obviously well aware of the issue and in fact offers a cloud version called Office 365.

Web-based services appeal to Microsoft and other vendors because it's easy to impose a subscription pricing model, where users have to pay an ongoing fee for access. Consumers, however, may prefer to buy the software once, and pay for upgrades if necessary. I'm still using a 2008 version of Microsoft Office on my older Macs. It works fine and (to Microsoft's credit) it still gets occasional security updates, so I'm in no hurry to upgrade.

Cloud computing relies on fast processing and lots of memory on the client, and high bandwidth to the server. Client side code is written in JavaScript and is usually intricate. The JavaScript code makes heavy demands on the browser to update and display graphical material quickly, responding to user actions like dragging and server actions like updated content. This is hard enough, but it's made worse by incompatibilities among browsers and versions of JavaScript, which require the supplier to figure out the best way to send the right code to the client. These are improving, however, as computers get faster and standards are more closely adhered to.

Cloud computing can trade off where computation is performed against where information resides during processing. For example, one way to make JavaScript code independent of the particular browser is to include tests in the code itself, like "if the browser is Firefox version 75 do this, else if it's Safari 12 do that, else do something different." Such code is bulky, which means that it requires more bandwidth to send the JavaScript program to the client, and the extra tests might make the browser run more slowly. As an alternative, the server can ask the client what browser is being used and then send code that has been tailored to that specific browser. That code is likely to be more compact and to run more quickly, though for a small program it might not make much difference.

Web page contents can be sent uncompressed, which requires less processing at both ends but more bandwidth; the alternative is to compress, which takes less bandwidth but requires processing on each end. Sometimes the compression is on one end only; large JavaScript programs are routinely squeezed by removing all unnecessary spaces and by using one- or two-letter names for variables and functions. The results are inscrutable to human readers but the client computer doesn't care.

In spite of technical challenges, cloud computing has much to offer, assuming you always have access to the Internet. Software is always up to date, information is stored on professionally managed servers with plenty of capacity, and client data is backed up all the time so there's little chance of losing anything. There's only one copy of a document, not multiple potentially inconsistent copies on different computers. It's easy to share documents and to collaborate on them in real time. The price is hard to beat, often free for individual consumers, though business customers may have to pay.

On the other hand, cloud computing raises difficult privacy and security questions. Who owns data stored in the cloud? Who has access to it and under what circumstances? Is there any liability if information is accidentally released? What happens to the accounts of people who have died? Who could force the release of data? For instance, in what situations would your email provider voluntarily or under threat of legal action release your correspondence to a government agency or to a court as part of a lawsuit? Would you find out if it did? In the US, companies can be forbidden, by a so-called "National Security Letter," from telling customers that they are the subject of a government request for information. How does the answer depend on where you live? What if you're a resident of the European Union, with its relatively strict rules about the privacy of personal data, but your cloud data is stored on servers in the US and subject to laws like the Patriot Act?

These are not hypothetical questions. As a university professor, I necessarily have access to private information about students—grades, of course, but occasionally sensitive personal and family information—that arrives by email and is stored on university computers. Would it be legal for me to use Microsoft's cloud services for my grade files and correspondence? If some accident on my part caused this information to be shared with the world, what could happen? What if Microsoft were subpoenaed by a government agency seeking information about a student or a bunch of them?

I'm not a lawyer and I don't know the answers, but I worry about this, so I try to avoid cloud services for student records and communications. I keep all such material on computers provided by the school, so if something private leaks out through their negligence or error, I should be somewhat protected from liability claims. Of course, if I screw up personally, it probably doesn't matter where the data was kept.

Who can read your mail and under what circumstances? This is partly a technical question and partly a legal one, and the answer to the legal part depends on what jurisdiction you live in. In the United States, my understanding is that if you are an employee of a company, your employer can read your mail on company accounts at will and without notice to you, period. That's true whether the mail is business related or not, on the theory that since the employer provides the facilities, it has the right to make sure that they are being used for business purposes and in accordance with company and legal requirements.

My mail is usually not very interesting, but it would bother me a lot if my employers were to read it without a serious reason even if they have the legal right to do so. If you're a student, most universities take the position that student email is private in the same way that student paper mail is. In my experience, students don't use their university mail accounts except as relays; they forward everything to Gmail. In tacit acknowledgment of this fact, many universities, including my own, outsource student mail to external services. These accounts are intended to be separate from the general services, subject to regulations about student privacy, and free of advertising, but the data still resides with the provider.

If you use an ISP or a cloud service for your personal mail, as most people do (for instance Gmail, Outlook, Yahoo and many others), privacy is between you and them. In general, such services take the public position that their customers' email is private and no human will look at it or reveal it short of legal requests, but they usually don't discuss how firmly they will resist subpoenas that seem too broad or informal requests that come cloaked as "national security." You're relying on the willingness of your provider to stand up to powerful pressures. The US government wants easier access to email, in the interests of fighting organized crime before 9/11 and terrorism after. The pressure for this kind of access increases steadily, and sharply after any terrorist incident.

As an example, in 2013, a small company called Lavabit that provided secure mail for its clients was ordered to install surveillance on the company network so the US government could access mail. The government also ordered the surrender of the encryption keys and told the owner, Ladar Levison, that he could not tell his clients that this was happening. Levison refused, arguing that he had been denied due process. Ultimately he chose to shut down his company rather than enable access to his customers' mail. It eventually became evident that the government was after information about a single account, that of Edward Snowden.

Today one might use ProtonMail as an alternative. It's based in Switzerland, promises privacy, and is certainly well-positioned to ignore requests for information that come from other countries. But any company can find itself squeezed by government agencies and commercial financial pressures, no matter where it is.

Leaving aside privacy and security concerns, what liability does Amazon or another cloud provider have? Suppose that some configuration error causes AWS service to be unacceptably slow for a day. What recourse does an AWS customer have? Service-level agreements are the standard way of spelling this out in contracts, but a contract doesn't ensure good service; it only provides a basis for legal action when something goes seriously wrong.

What responsibilities does a service provider have to its customers? When will it stand up and fight, and when is it likely to yield to legal threats or quiet requests from "the authorities"? There's no limit to the number of such questions, and few clear answers. Governments and individuals are always going to want more information about others while trying to curtail the amount of information available about themselves. A number of big players, including Amazon, Facebook and Google, now publish "transparency reports" that give a rough count of government requests to remove information, provide information about users, take down a copyright violation, and similar actions. Among other things, these reports give tantalizing hints of how often major commercial entities push back and on what grounds. For instance, in 2019 Google received more than 160,000 requests by governments for information on about 350,000 user accounts. It disclosed "some information" in about 70 percent of those. Facebook reports similar numbers of requests and disclosures.

11.6 Summary

We create voluminous and detailed streams of data as we use technology, much larger than we think. It's all captured for commercial use: shared, combined, studied, and sold, far more than we realize. This is the quid pro quo for the valuable free services that we take for granted, like search, social networks, phone apps and unlimited online storage. There is increasing public awareness (though not nearly enough) about the extent of data collection. Ad blockers are now being used by enough people that advertisers are starting to take notice. Given that advertising networks are often inadvertent suppliers of malware, blocking advertisements is prudent, but it's not clear what would happen if everyone started using Ghostery and Adblock Plus. Would the web as we know it come to a halt, or would someone invent alternative business models to support Google, Facebook and Twitter?

Data is also captured for government use, which seems more pernicious in the long run. Governments have powers that commercial enterprises do not and are accordingly harder to resist. How one tries to change governmental behavior varies greatly from country to country, but in all cases, being informed is a good first step.

A very effective AT&T advertising slogan of the early 1980s said "Reach out and touch someone." The web, email, texting, social networks, and cloud computing all make that easy. Sometimes this is just fine; you can make friends and share interests in a far larger community than you could ever meet in person. At the same time, the act of reaching out makes you visible and accessible throughout the world, where not everyone has your best interests at heart. That opens the door to spam, scams, spyware, viruses, tracking, surveillance, identity theft, and loss of your privacy and even your money. It's wise to be wary.

12

Artificial Intelligence and Machine Learning

"If a computer thinks, learns, and creates, it will be by virtue of a program that endows it with these capacities. [...] It will be a program that analyzes, by some means, its own performance, diagnoses its failures, and makes changes that enhance its future effectiveness."

Herbert A. Simon, *The New Science of Management Decision*, 1960.

"My colleagues, they study artificial intelligence; me, I study natural stupidity."
Amos Tversky (1937–1996), a founder of modern behavioral economics; quoted in *Nature*, April 2019.

If we apply steadily increasing computing power and memory to a huge amount of data, and stir in some sophisticated mathematics, it becomes possible to attack many long-standing problems of artificial intelligence: getting computers to behave in ways that we would normally think of as the sole province of humans.

Effective artificial intelligence is comparatively new, though there are historical roots going back to the 1950s. Today the field is an amalgam of buzzwords, hype, wishful thinking, and quite a few real accomplishments. Artificial intelligence, machine learning, and natural language processing (AI, ML, NLP) have been very successful for games (programs that play chess and Go are better than the best human players), speech recognition and synthesis (think Alexa and Siri), machine translation, image recognition and computer vision, and robotics systems like self-driving cars. Recommendation systems like those used by Netflix and Goodreads aim people at new movies and books, and Amazon's listings of related items surely contribute to the company's bottom line. Spam detection systems do a decent job, though keeping up with spammer techniques is an endless task.

Image recognition systems are remarkably effective at isolating components of pictures, then figuring out what those are, though they are often fooled. Medical image processing for identifying cancerous cells, retinal disease, and the like is sometimes as good as average clinicians, though not yet as good as the most expert. Face recognition is good enough for unlocking phones and doors, though it can be (and often is) abused for both commercial and government purposes.

There's a lot of jargon in the field and distinctly different things are sometimes blurred together. To set the stage, here are some brief explanations of terminology.

Artificial intelligence is the broad category of using a computer to do something that we normally think that only humans can do: the "intelligence" is what we credit ourselves with, and "artificial" means that a computer is doing it too.

Machine learning is a subset of artificial intelligence, a large family of techniques that are used to train algorithms so that they can make their own decisions and thus perform some of the tasks that we call AI.

Machine learning is not the same as statistics, though there is overlap. To greatly over-simplify, in a statistical analysis, we assume a model of the mechanism that produced some data, then try to find the parameters of that model that best fit the data. By contrast, a machine learning system does not assume a model, but tries to find relationships in the data. Normally ML systems are applied to larger datasets. Both statistics and machine learning are probabilistic: they give answers that have some probability of being correct, but offer no guarantees.

Deep learning is a specific form of machine learning that uses computational models that, at least metaphorically, are similar to the neural networks in our own brains. Deep learning implementations (very!) loosely mimic the kind of processing that the human brain appears to do: a set of neurons detects low-level features; their outputs provide inputs to other neurons that recognize higher-level features based on the lower level; and so on. As the system learns, some connections are strengthened while others are weakened.

Deep learning has been a highly productive approach, especially for computer vision. It is one of the most active areas of ML research, and there are a large number of different models.

There are zillions of books, scientific papers, popular articles, blogs and tutorials on these topics, and it's hard to keep up, even if one does nothing else in life. This chapter is a quick overview of machine learning that I hope will help you to understand some of the terminology, what ML is used for, how major systems work, how well they work, and, importantly, where they may fail.

12.1 Historical Background

Early in the development of computers in the mid-20th century, people began to think about how computers could be used to perform tasks that normally could only be done by a human being. One obvious target was playing games like checkers and chess, which have the advantage that the rules are completely specified, and there is a large population of humans who are interested and qualified as experts. Another target, translation from one language to another, was clearly more difficult, but more important. For example, machine translation from Russian to English was a critical matter during the Cold War era. Other applications included speech recognition and generation, mathematical and logical reasoning, decision making, and learning processes.

Funding for research on these topics was readily available, often from government agencies like the US Department of Defense. We've already seen the value of DoD funding for the early networking research that led to the development of the Internet. Research in artificial intelligence was similarly motivated and generously supported.

I think that it's fair to characterize AI research during the 1950s and 1960s as naively optimistic. Scientists felt that breakthroughs were right around the corner. In another five or ten years, computers would be translating languages accurately, and winning chess matches against the best human players.

I was just an undergraduate at the time, but I found the area and the potential results fascinating, and I wrote my senior thesis on artificial intelligence. Sadly, I no longer have a copy—it went missing somewhere along the way—and I don't recall how much of the prevailing optimism that I shared.

But almost every AI application proved to be much harder than was thought, and "another five or ten years" was always true. Results were sparse, funding ran out, and the field lay fallow for a decade or two, a period later referred to as the "AI winter." Then in the 1980s and 1990s work began on a different approach, *expert* or *rule-based* systems.

In an expert system, domain experts write down lots of rules, programmers convert the rules into code, and then computers apply them to perform some task. Medical diagnosis was one popular application area. Doctors would create rules for deciding what was wrong with a patient, and then programs could perform diagnoses, to support, complement, or in theory even replace real doctors. For example, MYCIN, one of the early examples, was designed to diagnose blood infections. It used about 600 rules, and was at least as good as general practitioners. MYCIN was developed by Edward Feigenbaum, one of the pioneers of expert systems; he shared the 1994 Turing Award for his work on artificial intelligence.

Expert systems had some real successes, including customer support, maintenance and repair of mechanical systems, and other focused areas, but it eventually became clear that there were major limitations as well. In practice, it's hard to collect a comprehensive set of rules, and there are too many exceptions. The technique doesn't scale well to large topics or to new problem domains. Rules need to be updated as conditions change or understanding improves. For example, think about how decision rules changed in 2020 for doctors who encountered a patient with an elevated temperature, a sore throat, and a bad cough. What might have once been the common cold, perhaps with minor complications, could easily be Covid-19, a highly contagious disease with serious risks for patients and health-care providers.

12.2 Classical Machine Learning

The basic idea of *machine learning* is to give an algorithm a large number of examples and let it "learn" for itself, without being given rules and without being explicitly programmed to solve specific problems. In the simplest form, we provide the program with a *training set* of examples that have been labeled with their correct values. For example, rather than trying to invent rules for how to recognize handwritten digits, we instead train a learning algorithm with a large number of sample digits, each labeled with its numeric value. The algorithm uses its successes and failures on the training data to learn how to combine features of the training data to get the best results. Of course, "best" is not a certainty: ML algorithms try to improve the probability of getting good results, but do not guarantee perfection.

After training, the algorithm classifies new items or predicts their values, based on what it learned from the training set.

Learning based on labeled data is called *supervised learning*. Most supervised learning algorithms have a common structure. They process a large number of examples that are labeled with the correct category, for example, whether some text is spam or not, or what kind of animal appears in a picture, or the likely price of a house. The algorithm determines parameter values that enable it to make the best classifications or predictions based on this training set. In effect, it learns how to generalize from examples.

We still have to tell the algorithm what "features" are relevant to making correct decisions, though not how to weight or combine those features. For instance, if we are trying to filter email, we need features that in some way relate to spam content, like spammy words ("free!"), known spam topics, weird characters, spelling mistakes, incorrect grammar, and so on. None of these features are individually definitive, but with enough tagged data, an algorithm can begin to separate spam from not-spam, at least until spammers adapt.

Handwritten digit recognition is a well-understood problem. NIST, the National Institute of Standards and Technology, provides a public test suite with 60,000 training images and 10,000 testing images; Figure 12.1 shows a small sample. Machine learning systems do well on this data: error rates in public competitions are below 0.25%, that is, about one mistake in 400 characters.

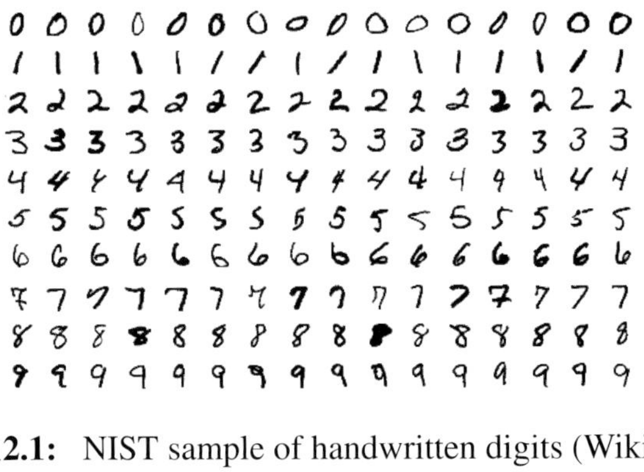

Figure 12.1: NIST sample of handwritten digits (Wikipedia).

ML algorithms can fail in many ways—for example, "over-fitting," in which the algorithm does well on its training data but much less well on new data. We may not have enough training data, or we may have the wrong set of features. Or the algorithm may produce results that confirm biases in the training data. This is an especially sensitive issue in criminal justice applications like sentencing or predicting recidivism, but also in any situation where an algorithm is making decisions about people: credit scoring, mortgage applications, and resumes.

Spam detection and digit recognition are examples of *classification*: putting items into the right class. *Prediction* algorithms try to predict some numeric value, like house prices, sports scores, or stock market trends. For example, we might try to predict house prices based on core features like location, age, living area and number of rooms; more complex models, such as those used by Zillow, would add features such as previous sale prices of similar houses, neighborhood characteristics, real estate taxes, and quality of local schools.

By contrast with supervised learning, *unsupervised learning* uses unlabeled training data, that is, data that isn't labeled or tagged with anything. An unsupervised learning algorithm tries to find patterns or structure in the data, grouping items based on their features. In one popular algorithm, k-means clustering, the algorithm does its best to assign the data to k groups in a way that maximizes similarities of items in the same group while minimizing similarities from one group to another. For example, to determine authorship of documents, we might hypothesize that there are two authors. We choose potentially relevant features like sentence length, vocabulary size, punctuation style, and so on, and let the clustering algorithm divide the documents into two groups as best it can.

Unsupervised learning is also useful for identifying outliers in some group of data items. If most items cluster in some obvious way but there are a handful that don't, they may represent data to be examined further. For example, suppose that the artificial data in Figure 12.2 represents some aspect of credit card usage. Most of the data points fall into one of the two big clusters, but a few points do not. They could be fine—the clustering need not be perfect—but they could also be instances of fraud or error.

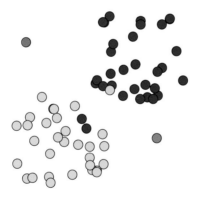

Figure 12.2: Clustering to identify anomalies.

Unsupervised learning has the advantage that there's no need for potentially expensive tagged or labeled training data, but it is not applicable to all situations. It requires figuring out a set of useful features that relate to clusters, and of course having a decent idea of how many clusters there might be. As a personal example, I once did an experiment to see what would happen if I used a standard k-means clustering algorithm to divide roughly 5,000 face images into two clusters. My naive hope was that this might group the population by gender. Empirically, it appeared to be correct 90 percent of the time, but I have no idea what it was basing its conclusions on, and I could see no obvious patterns in the errors.

12.3 Neural Networks and Deep Learning

If computers could simulate how the human brain works, they could perform as well as humans on intellectual tasks. That's sort of a holy grail for artificial intelligence, and people have tried that approach for many years.

Brain function is based on connections of neurons, which are special kinds of cells that respond to stimuli like touch, sound, light or inputs from other neurons. When its input stimuli are strong enough, a neuron "fires" and sends a signal on its outputs, which in turn may cause other neurons to react. (This is of course grossly over-simplified.)

Computer neural networks are simplified versions of such connections based on artificial neurons connected in regular patterns, as sketched schematically in Figure 12.3. Each neuron has a rule for how it combines its inputs, and each edge has a weight that is applied to data that passes along it.

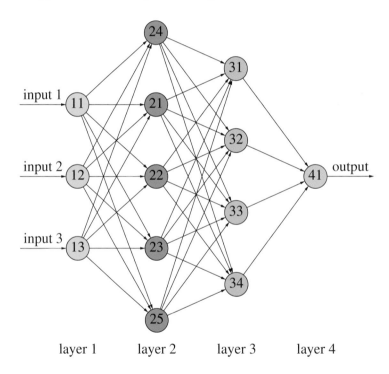

Figure 12.3: Artificial neural network with 4 layers.

Neural networks are not a new idea, but early work with them did not produce enough useful results, and they fell out of favor. Nevertheless, in the 1980s and 1990s, a handful of researchers continued to study them, and contrary to expectations, by the early 2000s, artificial neural networks were producing better results on tasks like image recognition than the best existing techniques. Most recent progress in machine learning is based on neural nets. The 2018 Turing Award was given to three of these persistent scientists: Yoshua Bengio, Geoffrey Hinton, and Yann LeCun.

The core idea of such networks is that early layers identify low-level features, for example recognizing patterns of pixels that might be edges in an image. Later layers identify higher-level features, like objects or regions of color, and ultimately the final layers identify entities like cats or faces. The word "deep" in deep learning refers to the fact that there are multiple layers of neurons; depending on the particular algorithm, this might be only a couple, or it might be a dozen or more.

The diagram in Figure 12.3 doesn't reveal the complexity of the computation done within the network, nor the fact that information flows backwards as well as forward, so that by iterating its processing and updating the weights at each node, the network improves its recognition at each layer.

The network learns by repeatedly processing the inputs and generating the output, for a very large number of iterations. At each iteration, the algorithm measures the error between what the neural network does and what we want it to do, and modifies the weights to try to reduce the error on the next iteration. It stops when we run out of training time, or the weights are not changing much.

One critical observation about neural networks is that they do not need to be given a set of features to look for. Rather, they find their own features, whatever those might happen to be, as part of the learning process. This leads to a potential downside of neural networks: they do not explain what "features" they have identified, and thus provide no specific explanation or understanding of their results. That's one reason why we must be cautious about blindly relying on them.

Deep learning has been especially successful for tasks related to computer vision, that is, getting computers to identify objects in images and sometimes to recognize specific instances, like human faces. For instance, Google Maps recognizes and blurs faces, license plates and sometimes house numbers in Street View. This is a much easier version of the general problem of recognizing a specific face, since there is little downside to blurring something that isn't a face.

Computer vision is central to a number of robotics applications, especially self-driving cars, which clearly have to be able to interpret the world around them, and do so quickly.

At the same time, face recognition in particular raises a large number of concerns. The potential for greatly increased surveillance is the most obvious, but face recognition enables subtle discrimination as well. Most face recognition systems do not work as accurately for people of color, because there is not enough diversity in the training images. In 2020, amid worldwide protests against racism, major companies announced plans to leave the field entirely (IBM) or to have a moratorium on supplying facial recognition technology to law enforcement (Amazon, Microsoft). None of these companies were big players in the area, so there was not much business effect, but perhaps the symbolism is a sign of deeper commitment.

One of the most dramatic successes of deep learning has been the creation of algorithms that play the most difficult human games, like chess and Go, better than the best human players. Not only are they better than humans, but they learned how to do this in a matter of hours by playing against themselves.

AlphaGo, a program developed by the company DeepMind (later acquired by Google) was the first program to defeat a professional Go player. It was quickly followed by AlphaGo Zero, which played substantially better, and then by AlphaZero, which played not only Go but also chess and the Japanese board game shogi, which is similarly difficult. AlphaZero taught itself how to play by playing games against itself, and with a day of training was able to beat the best conventional chess program, Stockfish, which it defeated in a 100-game match with 28 wins, no losses and 72 draws.

AlphaZero is based on a form of deep learning called *reinforcement learning*, which uses feedback from the external environment (in the case of games, whether it wins or loses) to continuously improve its performance. It needs no training data, because the environment tells it whether it's doing the right thing or at least heading in the right direction.

If you want to do some of your own experiments with machine learning, Google's teachablemachine.withgoogle.com makes it easy to experiment with tasks like image and sound recognition.

12.4 Natural Language Processing

Natural language processing (NLP) is a sub-area of machine learning that deals with how to make computers process human languages—how to understand what some text means, summarize it, translate it into another language, convert it into speech (or convert speech into text), or even generate meaningful text that appears to have come from a human. Today we see NLP in action with voice-actuated systems like Siri and Alexa, which recognize speech and turn it into text, figure out what the question is, search for relevant answers, and then synthesize a natural-sounding response.

Can a computer tell us what a document is about, what it means, or how it is germane to what we are working on? Can a computer create an accurate summary or synopsis of a long document? Can a computer find relevant or related texts, like the same news story from different perspectives, or legal cases that might be germane? Can it reliably detect plagiarism? These are all questions that could be addressed if computers were better at understanding language.

Sentiment analysis is one interesting special case of understanding, trying to determine whether a piece of text is fundamentally positive or negative. This has applications in assessing customer feedback and reviews, survey responses, news stories, and the like.

Naively, one might suppose that it would be sufficient to count the positive words ("wonderful", "great", "stable", "genius") and the negative words ("awful", "dumb") and display the difference. But this approach doesn't begin to capture the subtlety of language. Even simple negations are challenging ("He's neither stable nor a genius") and irony or sarcasm ("He's hardly wonderful") is much harder.

Another AI challenge is to carry on a conversation at a level that appears to be human. This harks back to Turing's test for computer intelligence, which we described at the end of Chapter 3. Intelligent conversation requires understanding what is said, and being able to generate appropriate responses.

One of the first attempts at using natural language to interact with a computer was a program called Eliza, which simulated a conversation between a patient and a Rogerian psychologist. Eliza was created by MIT professor Joseph Weizenbaum in

1966, and was surprisingly effective, at least when first encountered. There are plenty of Eliza simulators on the web; it's entertaining to play with them for a few minutes, though their weaknesses quickly become evident. Figure 12.4 is a sample conversation with one.

```
ELIZA: How do you do.  Please tell me your problem.
YOU:   Writing a book is hard work.
ELIZA: What does that suggest to you?
YOU:   Maybe I should do something different?
ELIZA: Can't you be more positive?
YOU:   Definitely I should do something different.
ELIZA: That's quite interesting.
YOU:   Is it?
ELIZA: That is interesting.  Please continue.
```

Figure 12.4: A "conversation" with Eliza.

Chatbots have improved since 1966, but again one doesn't need a lot of experience to discover their limitations. How often have you called some company and gotten into a dialog with a recorded voice that goes like this?

"In a few words, tell me how I can help you. You can say billing, repairs, orders, or 'something else.'"

Speech recognition software looks for keywords in your response, then sends you to the next step in this frustrating process. There's no real understanding, however, so if your problem doesn't match one of the keywords, it doesn't converge. Most times I don't even try, and attempt to bypass the whole sequence by repeatedly pushing zero. That generally gets me to a real person, who is much more likely to be helpful.

More sophisticated systems, based on language models trained on a large body of text, can generate better output, though not in response to specific human interactions. One interesting example is Talk to Transformer, which generates plausible continuations of an initial phrase provided by the user. It's based on a database of 8 million web pages; it tries to predict the next words based on patterns derived from all previous words. The result is usually at least plausible, and sometimes very good. Here are two results, where I provided the initial sentence "Writing a book is hard work."

Writing a book is hard work. Try to come up with something that makes sense, be unique, and that has your name and your vision on it. The copyright holder can argue that you're copying their book without their permission and even going as far as filing a lawsuit. You have to be able to handle all that and still write the best book possible.

Writing a book is hard work. And it's important to do this right. So many writers don't get it. I spent years of my life trying to produce The Book I Want To Write. Over the last year, my new book, Getting It Right the First Time, has been a passionate success. I've received notes from my editors and readers with gratitude and encouragement.

Through a strange combination of circumstances, in the spring of 2020 I co-taught a course on poetry and computation. One of the questions we debated was whether computers could write poetry as well as humans could, a kind of Turing test in a narrow domain. The answer depends on any number of ill-defined terms, so I can't say definitively, but for fun Figure 12.5 contains three poems for you to assess: human or program? The authors are revealed at the end of this chapter.

> Illegibility of this
> World. All twice-over.
> Robust Clocks
> agree the Cracked-Hour,
> hoarsely.
> You, clamped in your Depths,
> climb out of yourself
> for ever.

> Listening to find
> she hides deep within her
> yet in mortal reach.

> WHAT was the use of not leaving it there where it would hang what was the use if there was no chance of ever seeing it come there and show that it was handsome and right in the way it showed it. The lesson is to learn that it does show it, that it shows it and that nothing, that there is nothing, that there is no more to do about it and just so much more is there plenty of reason for making an exchange.

Figure 12.5: Three poems—which were written by program and which by human?

Computer translation of one human language into another is an old problem. Back in the 1950s, people confidently predicted that it would be a solved problem in the 1960s, and in the 1960s the expectation was the 1970s. Sadly, we're not there yet, though the situation is enormously better than it was, thanks to lots of computing power and large collections of text that can be used to train machine-learning algorithms.

The classic challenge is to translate the English expression "the spirit is willing but the flesh is weak" into Russian, then back to English. At least in legend, this came out as "the vodka is strong but the meat is rotten." Today, Google Translate provides the sequence seen in Figure 12.6. It's better but hardly perfect, a clear indication that machine translation is not yet a solved problem. (Google's algorithm changes frequently, so your results may well differ by the time you try it.)

Today machine translation is useful for getting a rough idea of what some text is about, especially if it is in a language or character set where you have no knowledge at all. But details are often wrong, and nuance escapes entirely.

| the spirit is willing but the flesh is weak. | × | дух желает, но плоть слаба. |
| дух желает, но плоть слаба. | × | the spirit desires, but the flesh is weak. |

Figure 12.6: Translation from English to Russian and back.

12.5 Summary

Machine learning is not a panacea, and there are many open questions about how well it works, and especially about how to explain the results that it comes up with. The xkcd cartoon in Figure 12.7 captures this perfectly.

Figure 12.7: xkcd.com/1838 on Machine Learning.

Artificial intelligence and machine learning have brought us breakthroughs in computer vision, speech recognition and generation, natural language processing, robotics, and many other areas. At the same time, they raise serious concerns about fairness, bias, accountability, and appropriate ethical uses for the technology. Perhaps the most important issue is that answers from ML systems may "look right" but only because they reflect biases in the data they started with.

It is possible to be led astray by artifacts of the training data. For instance, there's an old story that one study did an excellent job of detecting tanks in training pictures, but failed badly in practice. The reason: most of the training pictures were taken on sunny days, so the algorithm had learned to recognize nice weather, not tanks.

Unfortunately, this is just an appealing urban legend, as documented by Gwern Branwen at www.gwern.net/Tanks. But there is a useful caution even if the story isn't true: be sure you're not being misled by some irrelevant artifact.

Can ML algorithms be better than their data? Amazon scrapped an internal tool that it used for recruiting, because it clearly showed bias against female applicants. Amazon's models assessed candidates by observing patterns in resumes submitted to the company over a 10-year period. But most of those applicants had been male, so the training data was not representative of the current applicant pool. The effect was that the system learned to prefer male applicants. In short, no AI or ML system can be better than its input data, and there is a serious probability that such systems will merely confirm the biases inherent in their data.

For example, computer vision systems are able to identify faces, sometimes with reasonably high accuracy. That can be used for positive good, like unlocking your phone or your office, but it can also be put to more troubling uses. Smart doorbell systems like Amazon Ring can monitor what's going on near your house, and send alerts to you and to the local police when something suspicious happens. If the system starts to flag people of color in a predominantly white neighborhood as "suspicious," however, that seems like mechanization of racism.

Concerns about such issues led Amazon to suspend police use of its Rekognition software in mid-2020; the action was taken at a time of widespread protests over police brutality and racial bias in the US. Shortly thereafter, several suits were filed against Clearview AI, the company that has created a face database from billions of web photos and makes that data available to law enforcement agencies. Clearview argues that collecting publicly available information is protected by the free-speech provisions of the First Amendment.

Computer vision systems are used in a variety of surveillance settings. What are the limits? Should a military system for locating potential terrorist leaders call in drone strikes when such a person is apparently identified? How far should we go in mechanizing decisions like this? More generally, how should we deal with machine learning in safety-critical systems like self-driving cars, autopilots, industrial control systems, and many others? When there is no deterministic behavior to review or audit, how do we ensure that there are no circumstances in which a model will choose a disastrous course of action, for example, suddenly accelerating a self-driving car, or firing a missile into a crowd?

Machine learning models are sometimes used in the criminal justice system to predict whether someone accused of a crime is at risk of recidivism; this can affect bail and sentencing decisions. The problem is that training data reflects the current situation, which may well reflect systemic injustice based on race, gender and other characteristics that are in the data. Unbiasing such data is a difficult problem.

In many ways, we are still in the early days of artificial intelligence and machine learning. It seems likely that the benefits will continue, but so will the drawbacks, and we will have to be vigilant about recognizing and controlling the latter.

The first and last poems of Figure 12.5 are by humans, respectively Paul Celan and Gertrude Stein; the middle one was generated by Ray Kurzweil's Cybernetic Poet program, accessed through botpoet.com.

13

Privacy and Security

"You have zero privacy anyway. Get over it."
Scott McNealy, CEO of Sun Microsystems, 1999.

"Technology has now enabled a type of ubiquitous surveillance that had previously been the province of only the most imaginative science fiction writers."
Glenn Greenwald, *No Place to Hide*, 2014.

Digital technology has brought us an enormous number of benefits, and our lives would be much poorer without it. At the same time, it has had a large negative effect on personal privacy and security, and (in my editorial opinion) this is getting worse. Some of the erosion of privacy is related to the Internet and the applications it supports, while some is merely a by-product of digital devices getting smaller, cheaper, and faster. Increased processing power, storage capacity and communications bandwidth combine to make it easy to capture and preserve personal information from many sources, aggregate and analyze it efficiently, and disseminate it widely, all at minimal expense.

One interpretation of *privacy* is the right and ability to keep aspects of one's personal life unknown to others. I don't want the government or the companies that I deal with to know everything about what I buy, who I communicate with, where I travel, what books I read, what entertainment I enjoy. All of those things are my private business, and they should be revealed to others only when I explicitly approve. It's not that I have deep embarrassing secrets, at least no more than the average person does, but I should be secure in the knowledge that my life and habits are not shared with others, and especially not shared with commercial interests that want to sell me more things nor with government agencies no matter how well-intentioned.

People sometimes say "I don't care; I have nothing to hide." This is naive and silly. Would you like any random person to be able to see your home address, phone numbers, tax returns, emails, credit reports, medical records, all the places you've walked or driven, who you've exchanged phone calls and texts with? Not likely, but with the possible exception of your tax returns and medical records, all of that information is potentially available to data brokers, who can sell it on to others.

Governments use the word *security* in the sense of "national security," that is, protecting the country as a whole against threats like terrorist attacks and the actions of other countries. Corporations use the word to refer to protection of their assets from criminals and other companies. For individuals, security is often lumped with privacy, since it's hard to feel safe and secure if most aspects of your personal life are widely known or easy to find out. The Internet in particular has had a significant effect on our individual personal security—more financial than physical—since it has made it easy for private information to be gathered from many places, and it has opened up our lives to electronic intruders.

If you care about your personal privacy and online security, it's imperative to be more tech-savvy than most people are. If you learn the basics, you'll be a great deal better off than your less-informed friends. In Chapter 10 we looked at specific ways to manage browsing and phone use. In this chapter, we will look at broader countermeasures that individuals can take to slow down the invasions of their privacy and improve their security. This is a big subject, however, so this is a sample, not the whole story.

13.1 Cryptography

Cryptography, the art of "secret writing," is in many ways our best defense against attacks on our privacy. Properly done, cryptography is wonderfully flexible and powerful. Unfortunately, good cryptography is also difficult and subtle, and all too often defeated by human error.

Cryptography has been used for thousands of years to exchange private information with other parties. Julius Caesar used a simple encryption scheme (coincidentally called the Caesar cipher) of shifting the letters in his secret messages over by three positions, so A became D, B became E, and so on. Thus the message "HI JULIUS" would be encoded as "KL MXOLXV." This algorithm lives on in a program called `rot13` that shifts by 13 places. It's used on newsgroups to hide spoilers and offensive material from accidental viewing, not for any cryptographic purpose. (You might think about why shifting by 13 is convenient for English text.)

Cryptography has a long history, often colorful and sometimes dangerous to those who thought encryption would keep their secrets safe. Mary, Queen of Scots, lost her head in 1587 because of bad cryptography. She was exchanging messages with conspirators who wanted to depose Elizabeth I and put Mary on the throne. The cryptographic system was cracked, and a man-in-the-middle attack exposed the plot and the names of the collaborators; their fates make beheading seem humane. Admiral Isoroku Yamamoto, commander-in-chief of the Japanese Combined Fleet, was killed in 1943 because the Japanese encryption system was not secure. American intelligence learned Yamamoto's flight plans, and American pilots were able to shoot down his plane. And it is claimed, though not universally accepted, that World War II was significantly shortened because the British were able to decode German military communications encrypted with the Enigma machine (Figure 13.1), using computing techniques and expertise from Alan Turing.

Figure 13.1: German Enigma cipher machine.

The basic idea of cryptography is that Alice and Bob want to exchange messages, keeping the contents private, even though adversaries can read the messages that are being exchanged. To do this, Alice and Bob need some kind of shared secret that can be used to scramble and unscramble messages, making them intelligible to each other but unintelligible to everyone else. This secret is called the *key*. In the Caesar cipher, for example, the key would be the distance that the alphabet is shifted, that is, three for replacing A by D, and so on. For a complicated mechanical encryption device like the Enigma, the key is a combination of the settings of several code wheels and the wiring of a set of plugs. For modern computer-based encryption systems, the key is a large secret number used by an algorithm that transforms the bits of a message in a way that is infeasible to unscramble without knowing the secret number.

Cryptographic algorithms can be attacked in a variety of ways. Frequency analysis—counting the occurrences of each symbol—works well against the Caesar cipher and the simple substitution ciphers of newspaper puzzles. To defend against frequency analysis, algorithms must arrange that all symbols are equally likely and there are no patterns to analyze in the encrypted form. Another attack might be to exploit known plaintext, that is, a message known to have been encrypted with the target key, or to exploit chosen plaintext, running chosen messages through the algorithm with the target key. A good algorithm has to be immune to all such attacks.

It must be assumed that the cryptographic system is known and perfectly understood by adversaries, so that all of the security resides in the key. The alternative, assuming that adversaries don't know what scheme is being used or how it works, is called *security by obscurity*, and it never works for long, if at all. Indeed, if someone tells you that their cryptosystem is perfectly secure but they won't tell you how it works, you can be certain that it's not secure.

Open development of cryptographic systems is vital. Cryptosystems need the experience of as many experts as possible, probing for vulnerabilities. Even then it's tough to be sure that systems work. Algorithmic weaknesses can show up long after initial development and analysis. Bugs occur in code, inserted accidentally or malevolently. In addition there can be conscious attempts to weaken cryptographic systems, which seems to have been the case when the NSA tried to define critical parameters of one random number generator used in an important cryptographic standard.

13.1.1 *Secret-key cryptography*

Two fundamentally different kinds of cryptographic systems are used today. The older is usually called *secret-key cryptography* or symmetric-key cryptography. "Symmetric-key" is more descriptive because the same key is used for encryption and decryption, but "secret-key" contrasts better with the name of the newer kind, *public-key cryptography*, which is covered in the next section.

In secret-key cryptography, a message is encrypted and decrypted with the same secret key, which must be shared by all the parties who want to exchange messages. Assuming that the algorithm is entirely understood and has no flaws or weaknesses, the only way to decrypt a message is a *brute force attack*: try all possible secret keys, looking for the one that was used for encryption. This can take a long time; if the key has N bits, the effort is roughly proportional to 2^N. Brute force does not mean dumb, however. An attacker will try short keys before long ones and likely keys before unlikely ones, for example a *dictionary attack* based on common words and number patterns like "password" and "12345." If people are lazy or careless about choosing keys, such attacks can be very successful.

Until the early 2000s, the most widely used secret-key cryptographic algorithm was the Data Encryption Standard, or DES, which was developed by IBM and the NSA in the early 1970s. There was some suspicion that the NSA had arranged a secret backdoor mechanism so that DES-encoded messages could be easily decoded, but this was never confirmed. In any case, DES had a 56-bit key, and as computers became faster, 56 bits proved to be too short; by 1999, a DES key could be cracked by brute force in a day of computing by a fairly inexpensive special-purpose computer. This led to the creation of new algorithms with longer keys.

The most widely used of these is the Advanced Encryption Standard, or AES, which was developed as part of a worldwide open competition sponsored by the US National Institute of Standards and Technology (NIST). Several dozen algorithms were submitted from all over the world and were subjected to intense public testing and critique. Rijndael, created by the Belgian cryptographers Joan Daemen and Vincent Rijmen, was the winner, and became an official US government standard in 2002. The algorithm is in the public domain and anyone can use it without licenses or fees. AES supports three key lengths—128, 192, and 256 bits—so there are a lot of potential keys and a brute force attack will not likely be feasible for many years unless some weakness is discovered.

We can even put numbers on this. If a specialized processor like a GPU can perform 10^{13} operations per second, a million GPUs could do 10^{19} operations per second, or about 3×10^{26} per year, which is roughly 2^{90}. That's a long way from 2^{128}, so even AES-128 should be safe from a brute-force attack.

The big problem with AES and other secret key systems is *key distribution*: each communicating party must know the key, so there must be a secure way to get the key to each of them. That could be as easy as having everyone over to your house for dinner, but if some participants are spies or dissidents in hostile environments, there might not be any safe and secure channel for sending a secret key. Another problem is *key proliferation*: to have independent secret conversations with unrelated parties, you need a separate key for each group, making the distribution issue even harder. Considerations like these led to the development of public-key cryptography, our next topic.

13.1.2 Public-key cryptography

Public-key cryptography is a completely different idea, invented in 1976 by Whitfield Diffie and Martin Hellman at Stanford, using some ideas from Ralph Merkle. Diffie and Hellman shared the 2015 Turing Award for this work. The idea had been independently discovered a few years earlier by James Ellis and Clifford Cocks, two cryptographers at the British intelligence agency GCHQ, but their work was kept secret until 1997, so they were unable to publish and thus missed out on most of the credit.

In a public-key cryptosystem, each person has a *key pair*, consisting of a public key and a private key. The keys of the pair are mathematically related and have the property that a message encrypted with one key of the pair can only be decrypted with the other key, and vice versa. If the keys are sufficiently long, it is computationally infeasible for an attacker to decrypt the secret message or deduce the private key from the public key. The best known algorithms that an attacker can use require run time that grows exponentially with the key length.

In use, the public key is truly public: it is available to everyone, often posted on a web page. The private key must be kept strictly private, a secret known only to the owner of this pair of keys.

Suppose that Alice wants to send a message to Bob, encrypted so that only Bob can read it. She goes to Bob's web page and gets his public key, which she uses to encrypt her message to him. When she sends the encrypted message, an eavesdropper Eve might be able to see that Alice is sending a message to Bob, but since it's encrypted, Eve can't tell what the message says.

Bob decrypts Alice's message with his private key, which only he knows, and which is the only way to decrypt a message that was encrypted with his public key. (See Figure 13.2.) If he wants to send an encrypted reply to Alice, he encrypts it with *Alice's* public key. Again, Eve can see the reply but only in its encrypted form, which she can't understand. Alice decrypts Bob's reply with her private key, which only she knows.

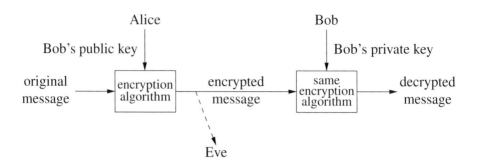

Figure 13.2:   Alice sends Bob an encrypted message.

This scheme solves the key distribution problem, since there are no shared secrets to distribute. Alice and Bob have their respective public keys on their web pages, and anyone can carry on a private conversation with either one of them without pre-arrangement or the exchange of any secrets. There is no need for the parties to have ever met. Of course, if Alice wants to send the same encrypted message to Bob, Carol, and others, she has to encrypt it separately for each recipient, using the right public key for each one.

Public-key cryptography is a critical component of secure communication on the Internet. Suppose that I want to buy a book online. I have to tell Amazon my credit card number, but I don't want to send that in the clear so we need an encrypted communication channel. Amazon and I can't use AES directly because we don't have a shared key. To arrange a shared key, my browser generates a random temporary key. It then uses Amazon's public key to encrypt the temporary key and sends it securely to Amazon. Amazon uses its private key to decrypt the temporary key. Amazon and my browser now use that shared temporary key to encrypt information like my credit card number with AES.

One drawback of public-key cryptography is that its algorithms tend to be slow, perhaps several orders of magnitude slower than a secret-key algorithm like AES. Thus, rather than encrypting everything with public-key, there's a two-step sequence: use public-key to agree on a temporary secret key and then use AES for transferring data in quantity.

The communication is secure at each stage, initially with public-key to set up the temporary key and then with AES for exchange of bulk data. If you visit a web store, an online mail service, and most other sites, you are using this technique. You can see it in action, because your browser will show that you are connecting with the HTTPS protocol (HTTP with security) and will display an icon of a closed lock to indicate that the link is encrypted:

Most web sites now use HTTPS by default. This might make transactions a bit slower but not by much, and the added security is important even if the specific use has no immediate reason to require a secure communication.

Public-key cryptography has other useful properties. For instance, it can be used to implement a *digital signature* scheme. Suppose Alice wants to sign a message so the recipient can be sure that it came from her and not from an impostor. If she encrypts the message with her private key and sends the result, then anyone can decrypt it with her public key. Assuming that Alice is the only one who knows her private key, the message must have been encrypted by Alice. Clearly this only works if Alice's private key has not been compromised.

You can also see how Alice could sign a private message to Bob, so that no one else can read it and Bob can be sure it came from Alice. Alice first signs the message to Bob with her own private key, then encrypts the result with Bob's public key. Eve can see that Alice sent something to Bob, but only Bob can decrypt it. He decrypts the outer message with his own private key, and then decrypts the inner message with Alice's public key to verify that it came from her.

Public-key cryptography doesn't solve everything, of course. If Alice's private key is revealed, all past messages to her can be read and all of her past signatures are suspect. It's hard to revoke a key, that is, to say that a particular key is no longer valid, though most key-creation schemes include information about when the key was created and when it is meant to expire. A technique called *forward secrecy* helps. Each individual message is encrypted with a one-time password as above and then the password is discarded. If the one-time passwords are generated in such a way that an adversary can't recreate them, then knowing the password for one message is of no help for decrypting previous or future messages even if the private key is compromised.

The most widely used public-key algorithm is called *RSA*, after Ronald Rivest, Adi Shamir and Leonard Adleman, the computer scientists who invented it at MIT in 1978. The RSA algorithm is based on the difficulty of factoring large composite numbers. RSA works by generating a large integer, at least 500 decimal digits long, that is the product of two large prime numbers each about half as long as the product, and uses these as the basis of the public and private keys. Someone who knows the factors (the holder of the private key) can quickly decrypt an encrypted message, while everyone else in effect has to factor the large integer, and this is believed to be computationally infeasible. Rivest, Shamir and Adleman won the 2002 Turing Award for their invention of the RSA algorithm.

The length of the key is important. So far as we know, the computational effort required to factor a big integer that is the product of two primes of about the same size grows rapidly with its length and factoring is infeasible. RSA Laboratories, the company that held the rights to the RSA patent, ran a factoring challenge from 1991 until 2007. It published a list of composite numbers of increasing lengths and offered cash prizes for the first person to factor each one. The smallest numbers were about 100 decimal digits long and were factored fairly quickly. When the challenge ended in 2007, the largest number factored had 193 digits (640 bits) and a prize of $20,000; in 2019, RSA-240 (240 digits, 795 bits) was factored. The list can still be found online if you want to try your hand.

Because public-key algorithms are slow, documents are often signed indirectly, using a much smaller value derived from the original in a way that can't be forged. This short value is called a *message digest* or *cryptographic hash*. It is created by an

algorithm that scrambles the bits of any input into a fixed-length sequence of bits—
the digest or hash—with the property that it's computationally infeasible to find
another input with the same digest. Furthermore, the slightest change to the input
changes approximately half the bits in the digest. Thus, any tampering with a document can be efficiently detected by comparing its digest or hash with the original
digest.

To illustrate, in ASCII the letters x and X differ by a single bit; in hex they are 78
and 58, and in binary 01111000 and 01011000. Here are their cryptographic
hashes using an algorithm called MD5. The first row is the first half of the hash of x
and the second row is the first half of the hash of X; the third and fourth rows are the
second halves. It's easy to count how many bits differ (66 of 128) by hand, though I
used a program.

```
10011101 11010100 11100100 01100001 00100110 10001100 10000000 00110100
00000010 00010010 10011011 10111000 01100001 00000110 00011101 00011010

11110101 11001000 01010110 01001110 00010101 01011100 01100111 10100110
00000101 00101100 01011001 00101110 00101101 11000110 10110011 10000011
```

It is computationally infeasible to find another input that has the same hash value as
either of these, and there's no way to go back from the hash to the original input.

Several message digest algorithms are in widespread use. MD5, illustrated above,
was developed by Ron Rivest; it produces a 128-bit result. SHA-1, from NIST, has
160 bits. Both MD5 and SHA-1 have been shown to have weaknesses and their use
is deprecated. SHA-2, a family of algorithms developed by the NSA, has no known
weaknesses. Nevertheless, NIST ran an open competition, analogous to the one that
produced AES, to create a new message digest algorithm; the winner, now known as
SHA-3, was selected in 2015. SHA-2 and SHA-3 provide a range of digest sizes
from 224 to 512 bits.

Although modern cryptography has amazing properties, in practice it still requires
the existence of trusted third parties. For example, when I order a book, how can I be
sure that I am talking to Amazon and not a sophisticated impostor? When I visit,
Amazon verifies its identity by sending me a *certificate*, a digitally signed collection
of information from an independent *certificate authority* that can be used to verify
Amazon's identity. My browser checks this with the certificate authority's public key
to verify that it belongs to Amazon and not someone else. In theory I can be sure that
if the authority says it's Amazon's certificate, it really is.

But I have to trust the certificate authority; if that's a fraud, then I can't trust anyone who uses it. In 2011 a hacker compromised DigiNotar, a certificate authority in
Holland, and was able to create fraudulent certificates for a number of sites, including
Google. If an impostor had sent me a certificate signed by DigiNotar, I would have
accepted the impostor as the real Google.

A typical browser knows about an astonishing number of certificate authorities—
nearly 80 in my version of Firefox and over 200 for Chrome—and the majority are
organizations that I've never heard of and are located in faraway places.

Let's Encrypt is a non-profit certificate authority that provides free certificates to
anyone, with the idea that if it's easy to get a certificate, eventually all web sites will
work with HTTPS, and all traffic will be encrypted. By early 2020, Let's Encrypt
had issued a billion certificates.

13.2 Anonymity

Using the Internet reveals a lot about you. At the lowest level, your IP address is a necessary part of every interaction, and that reveals your ISP and lets anyone make a guess about where you are. Depending on how you connect to the Internet, that guess might be accurate if you're a student at a small college, for example, or unrevealing if you're inside a large corporate network.

Using a browser, which is the most common case for most people, more is revealed (Figure 11.3). The browser sends the URL of the referring page and detailed information about what kind of browser it is and what kinds of responses it can handle (compressed data or not, for example, or what kinds of images it will accept). With suitable JavaScript code, browsers will report what fonts are loaded and other properties that, taken together, may make it possible to identify the specific user within a population of literally millions. This kind of browser fingerprinting is becoming common, and it's hard to defeat.

As we saw in Chapter 11, `panopticlick.eff.org` lets you estimate just how unique you are. In my experiments with one laptop, I am unique among over 280,000 recent users when I access the site with Chrome. One other person has the same Firefox settings as I do, and one other when I use Safari. These values vary depending on defenses like ad blockers, but most of the discrimination comes from the User Agent header (Figure 11.3) that is sent automatically and the fonts and plugins that have been installed, over which I have almost no control. Browser suppliers could send less of this potential tracking information, but little seems to have been done to improve the situation. Somewhat discouragingly, if I disable cookies or enable Do Not Track, that makes me a little more distinctive and thus easier to identify specifically.

Some web sites promise anonymity. For instance, Snapchat users can send messages, pictures and videos to friends, with the promise that the content will disappear within a specified short time. How much can Snapchat resist the threat of legal action? Snapchat's privacy policy says "We may share information about you if we reasonably believe that disclosing the information is needed to comply with any valid legal process, governmental request, or applicable law, rule, or regulation." This kind of language is common in all privacy policies, of course, and suggests that your anonymity is not very strong, and it will vary depending on what country you are in.

13.2.1 Tor and the Tor Browser

Suppose you're a whistleblower who wants to publicize some malfeasance without being identified. (Think Edward Snowden.) What if you're a dissident in a repressive regime, or a gay person in a country where gays are persecuted, or perhaps an adherent of the wrong religion? Or maybe, like me, you just want to use the Internet without being watched all the time. What can you do to make yourself less identifiable? The suggestions at the end of Chapter 10 will help, but one other technique is effective, though not without modest cost.

It's possible to use cryptography to conceal a conversation sufficiently well that the ultimate recipient of a connection does not know where that connection originated. The most widely used such system is called *Tor*, which originally stood for

"The Onion Router," a metaphorical play on the layers of encryption that surround a conversation as it passes from one place to another. The Tor logo in Figure 13.3 hints at the etymology.

Figure 13.3: The Tor logo.

Tor uses cryptography to send Internet traffic through a series of relays so that each relay only knows the identities of the immediately adjacent relays on the path and no others. The first relay on the path knows who the originator is but not the ultimate destination. The last relay on the path (the "exit node") knows the destination but does not know who originated the connection. A relay in the middle knows only the relay that provided the information and the relay to which it is being sent, but no more. The actual content is encrypted at every step.

The message is wrapped in multiple layers of encryption, one for each relay. Each relay removes one layer as it sends the message onward (hence the onion metaphor). The same technique is used in the reverse direction. Normally three relays are used, so the one in the middle knows nothing about the origin or destination.

At any given time, there are about 7,000 relays worldwide. The Tor application picks a random set of relays and sets up the path, which is changed from time to time, even during a single session.

The most common way to use Tor is through the Tor Browser, a version of Firefox that has been configured to use Tor for transport; it also sets Firefox privacy settings appropriately. Download it from `torproject.org`, install it and use it like any other browser, though pay attention to the warnings about how to use it securely.

The browsing experience is pretty much identical to Firefox, though perhaps a bit slower because it does take time to go through the extra routers and layers of encryption. Some web sites also discriminate against Tor users, sometimes in self-defense, since attackers appreciate anonymity as much as good people do.

As an example of how anonymization might appear to a casual user, Figure 13.4 is a view of the weather in Princeton as seen from Tor (on the left) and Firefox (on the right). For each, I visited `weather.yahoo.com`. Yahoo thinks it knows where I am, but it's wrong when I use Tor. Almost every time I have tried the experiment, the exit node is somewhere in Europe; reloading the page an hour later moved me from Latvia to Luxembourg. The only thing that gives me a bit of pause is that the temperatures are always reported in Fahrenheit, which is not much used outside of the US. How did Yahoo decide? Other weather sites do report in Celsius.

According to Panopticlick, when I use the Tor browser, about 3,200 other people in their sample of 280,000 recent visitors have the same characteristics as I do, so I'm

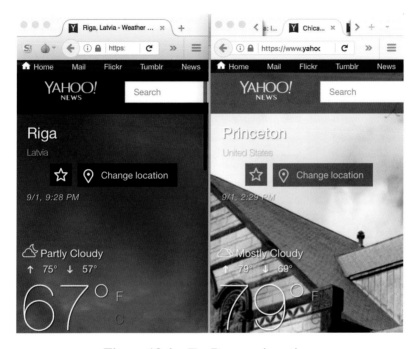

Figure 13.4: Tor Browser in action.

harder to identify by browser fingerprinting and certainly less distinctive than when using a direct browser connection. That said, Tor is by no means a perfect solution to all privacy issues. If you use it carelessly your anonymity could be compromised. Browsers and exit nodes can be attacked and a compromised relay would be a problem. It's also true that if you're using Tor you will stand out in a crowd, which may be a problem, though that gets better as more people use Tor.

Is Tor secure from the NSA or a similarly capable organization? One of the Snowden documents was a 2007 NSA presentation in which one slide (Figure 13.5) says "We will never be able to de-anonymize all Tor users all the time." Of course the NSA won't just give up, but so far Tor seems to be the best privacy tool that ordinary folk can use. (It is mildly ironic that Tor was originally developed by a US government agency, the Naval Research Laboratory, to help secure US intelligence communications.)

If you're feeling especially paranoid, try a system called TAILS, "The Amnesic Incognito Live System," a flavor of Linux that runs from a bootable device like a DVD, a USB drive, or an SD card. It runs Tor and the Tor browser, and it leaves no trace on the computer that it runs on. Software running under TAILS uses Tor for connecting to the Internet, so you should be anonymous. It also stores nothing on local secondary storage, just in primary memory; when the computer is shut down after a TAILS session, the memory is cleared. That lets you work on documents without leaving any record on the host computer. TAILS also provides a suite of other cryptographic tools, including OpenPGP, which lets you encrypt mail, files, and anything else. TAILS is open source and can be downloaded from the web.

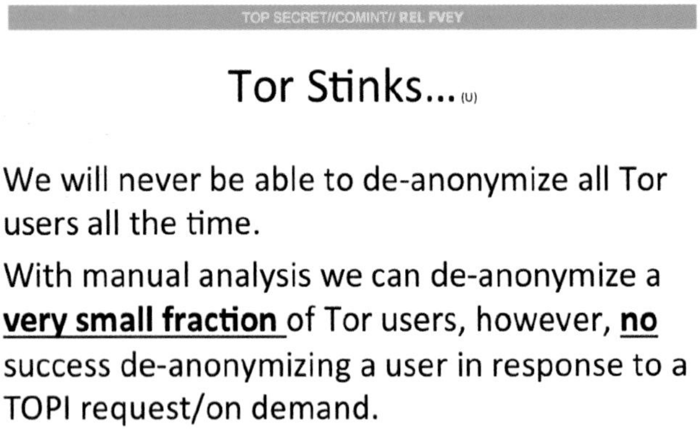

Tor Stinks... (U)

- We will never be able to de-anonymize all Tor users all the time.
- With manual analysis we can de-anonymize a **very small fraction** of Tor users, however, **no** success de-anonymizing a user in response to a TOPI request/on demand.

Figure 13.5: NSA presentation on Tor (2007).

13.2.2 Bitcoin

Sending and receiving money is another domain where anonymity is valued highly. Cash is anonymous: if you pay in cash, there's no record and no way to identify the parties involved. It's increasingly difficult to use cash today except for small local purchases like gas and groceries. Car rentals, plane tickets, hotels, and of course online shopping all require the use of credit or debit cards, which identify the purchaser. Plastic is convenient, but when you use a card or shop online, you leave a trail.

It turns out that clever cryptography can be used to create anonymous currency. The most successful example is called *Bitcoin*, a scheme invented by Satoshi Nakamoto and released as open-source software in 2009. (Nakamoto's real identity is unknown, an unusual example of successful anonymity.)

Bitcoin is a decentralized digital currency or cryptocurrency; it is not issued or controlled by any government or other party and it has no physical form, unlike the bills and coins of conventional money. Its value does not come from fiat, as money issued by governments does, nor is it based on some scarce natural resource like gold. Like gold, however, its value depends on how much of it users are willing to pay or accept for goods and services.

Bitcoin uses a peer-to-peer protocol that lets two parties exchange *bitcoins* without using an intermediary or trusted third party, in a way that emulates cash. The Bitcoin protocol ensures that the bitcoins are truly exchanged, that is, ownership transfers, no coins are created or lost in the transaction, and the transaction can't be reversed, yet the parties can remain anonymous both to each other and to the rest of the world.

Bitcoin maintains a *public ledger* of all transactions, called the *blockchain*, though the parties behind transactions are anonymous, identified only by an address that is in effect a cryptographic public key. Bitcoins are created ("mined") by doing a certain amount of computationally difficult work to verify and store payment information in the public ledger. Blocks in the blockchain are digitally signed and refer back to earlier blocks, so earlier transactions can't be modified without redoing all the work that

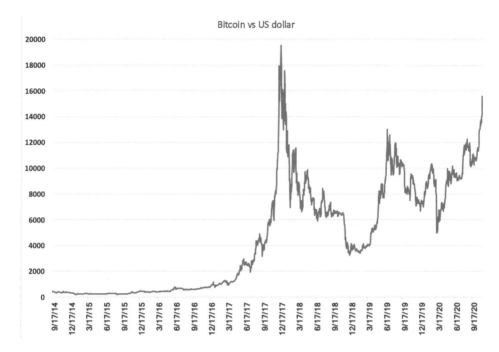

Figure 13.6: Bitcoin prices (finance.yahoo.com).

went into creating the blocks originally. Thus the state of all transactions from the beginning of time is implicit in the blockchain and could in principle be recreated. No one could fake a new blockchain without redoing all the work, which would be computationally infeasible.

It's important to note that the blockchain is totally public. Thus Bitcoin anonymity is more like "pseudonymity": everyone knows everything about all transactions associated with a specific address but they don't know that the address is yours. However, you could become linked to your transactions if you do not manage your addresses properly. You could also lose your bitcoins forever if you lose a private key.

Because the parties behind transactions can remain anonymous if they are careful, Bitcoin is a favored currency for drug deals, ransomware payoffs and other illegal activities. An online marketplace called Silk Road was widely used for illegal drug sales, paid for in bitcoins. Its proprietor was eventually identified not because of any flaws in anonymity software, but because he had left a sparse trail of online comments that a diligent IRS agent was able to track back to a real-world identity. Operations security ("opsec" in intelligence jargon) is difficult to do right, and it only takes one slipup to give the game away.

Bitcoins are a "virtual currency" but they can be converted to and from conventional currency. Historical exchange rates have been volatile; a bitcoin's value against the US dollar has wandered up and down by significant factors. Figure 13.6 shows prices for a multi-year period.

Big players like banks and even companies like Facebook have dipped their toes into the cryptocurrency world, by offering services or even their own version of a blockchain currency. Tax authorities are also interested in bitcoins, of course, since one use of anonymous exchanges is to escape taxation. In the US, virtual currencies like Bitcoin are treated as property for federal income tax purposes, and thus there can be taxable capital gains on transactions.

It's easy to experiment with bitcoin technology; `bitcoin.org` is a good place to start, and `coindesk.com` has excellent tutorial information. There are also books and online courses.

13.3 Summary

Cryptography is a vital part of modern technology; it's the basic mechanism that protects our privacy and security as we use the Internet. The unfortunate fact, however, is that cryptography helps everyone, not just the good guys. That means that criminals, terrorists, child pornographers, drug cartels and governments are all going to use cryptographic techniques to further their interests at the expense of yours.

There's no way to put the cryptographic genie back in the bottle. World-class cryptographers are few in number and scattered all over; no country has a monopoly on them. Furthermore, cryptographic code is mostly open source, available to anyone. Thus trying to outlaw strong cryptography in any given country is unlikely to prevent its use.

There are regular heated debates about whether encryption technology helps terrorists and criminals and thus should be outlawed or, more realistically, cryptosystems should have a "backdoor" through which suitably authorized government agencies can decrypt whatever the adversaries have encrypted.

Experts uniformly believe that this is a bad idea. In 2015, one especially qualified group published a report called "Keys under doormats: Mandating insecurity by requiring government access to all data and communications"; the title hints at their considered opinion.

Cryptography is exceedingly difficult to get right in the first place; adding intentional weaknesses, however carefully designed, is a recipe for larger failures. As we have seen repeatedly, governments (mine and yours) are terrible at keeping secrets— think about Snowden and the NSA. So relying on a government agency to keep backdoor keys safe and used appropriately is *a priori* a bad idea, even assuming good faith, which is a big assumption.

The fundamental problem is that we can't weaken the encryption that we want terrorists to use without weakening it for everyone. As Tim Cook, the CEO of Apple, said, "The reality is, if you put a backdoor in, that backdoor's for everybody, for good guys and bad guys." And of course crooks, terrorists, and other governments won't use the weakened version anyway, so we wind up worse off.

Apple software encrypts all the contents of iPhones running iOS, using a key provided by the user and unknown to Apple. If a government agency or a judge tells Apple to decrypt the phone, Apple can truthfully say that it is not able to do so. Apple's stand has not won it friends among politicians or law enforcement, but it is a defensible position. It also makes commercial sense, since savvy customers would be reluctant to buy phones where government agencies could easily snoop on contents and conversations.

In late 2015, two terrorists killed 14 people in San Bernardino, California, before being killed themselves. The FBI tried to force Apple to break the encryption on one of the terrorist's iPhones. Apple contended that creating even a special-purpose mechanism to access the information would create a precedent that would gravely weaken all phone security.

The San Bernardino incident eventually became moot when the FBI claimed to have found an alternate way to recover the information, but the issue returned after another shooting in Florida in late 2019. The FBI requested help. Apple says that it has provided all the information that it has; it does not have passwords.

The debate is intense, with both sides having valid concerns. My personal position is that strong encryption is one of the few defenses that ordinary people have against government over-reaching and criminal invasion, and we must not give it up. As noted when we talked about metadata earlier, there are plenty of other ways in which law enforcement agencies can obtain information, requiring only that a decent case be made. It should not be necessary to weaken everyone's encryption to investigate a small number of people. These are difficult issues, however, and they often arise in politically and emotionally charged situations like the aftermath of some violent event. We are not likely to see a satisfactory resolution in the short run.

In any security system, the weakest link is the people involved, who will accidentally or intentionally subvert systems that are too complicated or hard to use. Think about what you do when you are forced to change your password, especially when the new one has to be invented right now and must satisfy weird requirements like both upper and lower case letters, at least one digit, and some special characters but not others. Most people resort to formulas and writing it down, both of which potentially compromise security. Ask yourself: if an adversary saw two or three of your passwords, could he or she guess others? Think about spear phishing. How many times have you gotten almost plausible email that asked you to click or download or open? Were you tempted?

Even if everyone tries hard to be secure, a determined adversary can always use the four B's (bribery, blackmail, burglary, brutality) to gain access. Governments can use the threat of jail for people who refuse to divulge their passwords when asked. Nevertheless, if you are careful, you can do a decent job of protecting yourself, not from all threats all the time, but enough that you can function in the modern world.

14

What Comes Next?

"Making predictions is hard, especially about the future."

Attributed to Yogi Berra, Niels Bohr, Samuel Goldwyn, and Mark Twain, among others.

"Teachers should prepare the student for the student's future, not for the teacher's past."

Richard Hamming, *The Art of Doing Science and Engineering: Learning to Learn*, 1996.

We've covered a lot of ground. What should you have learned along the way? What's likely to matter in the future? What computing issues will we still be wrestling with in five or ten years? What will have become passé or irrelevant?

Superficial details change all the time, and many of the technical minutiae that I've talked about aren't terribly important except as a concrete way to help you understand how things work—most people learn better from specific instances than from abstractions, and computing has altogether too many abstract ideas.

On the hardware side, it's helpful to understand how computers are organized, how they represent and process information, what some of the jargon and numbers mean, and how they have changed over time.

For software, it's important to know how one defines computational processes precisely, both as abstract algorithms (with some sense of how their computation time grows with the amount of data) and as concrete computer programs. Knowing how software systems are organized, how they are created with programs in a variety of languages and often built from components, helps you to understand what is behind the major pieces of software that we all use. With luck, the bit of programming in a couple of chapters is enough that you could reasonably think about writing more code yourself, and even if you never do so, it's good to know what's involved.

Communications systems operate both locally and across the world. It's important to understand how information flows, who has access to it, and how it's all controlled. Protocols—rules for how systems interact—are important as well, since their properties can have profound effects, as seen in the problems of authentication in the Internet today.

Some computing ideas are useful ways of thinking about the world. For example, I have frequently made the distinction between logical structure and physical implementation. This central idea shows up in myriad guises. Computers are a fine example: how computers are constructed changes rapidly but the architecture has remained much the same for a long time. More generally, digital computers all have the same logical properties—in principle they can all compute the same things. In software, code provides an abstraction that hides implementation; implementations can be changed without changing things that use them. Virtual machines, virtual operating systems, and indeed even real operating systems are all examples of the use of interfaces to separate logical structure from actual implementation. Arguably, programming languages also serve this function, since they make it possible for us to talk to computers as if they all spoke the same language and it was one that we understood too.

Computer systems are good examples of engineering tradeoffs, a reminder that one never gets something for nothing—there is no free lunch. As we saw, a desktop, a laptop, a tablet and a cell phone are equivalent computing devices, but they differ markedly in how they deal with constraints on size, weight, power consumption, and cost.

Computer systems are also good examples of how to divide large and complicated systems into smaller and more manageable pieces that can be created independently. Layering of software, APIs, protocols and standards are all illustrations.

The four "universals" that I mentioned in the introduction will remain important for understanding digital technology. To recapitulate:

First is the *universal digital representation of information*. Chemistry has more than 100 elements. Physics has well over a dozen elementary particles. Digital computers have two elements, zero and one, and everything else is built of those. Bits represent arbitrary kinds of information, from the simplest binary choices like true and false or yes and no, through numbers and letters, to anything at all. Large entities—say the record of your life derived from your browsing and shopping, your phone calls, and ubiquitous surveillance cameras—are collections of simpler data items, and so on, down to the level of individual bits.

Second is the *universal digital processor*. A computer is a digital device that manipulates bits. The instructions that tell the processor what to do are encoded as bits and are generally stored in the same memory as data; changing the instructions causes the computer to do something different, which is the reason why computers are general-purpose machines. The meaning of bits depends on context—one person's instructions are another person's data. Processes like copying, encryption, compression and error detection can be performed on bits independent of what they represent, though specific techniques may work better on known kinds of data. The trend to replacing specialized devices with general-purpose computers that run general-purpose operating systems will continue. The future may well bring other kinds of processors based on biological computing or quantum computers or something yet to be invented, but digital computers will be with us for a long while.

Third is the *universal digital network*, which carries the bits, both data and instructions, from one processor to another, anywhere in the world. It's likely that the Internet and the telephone network will blend together into a universal network, mimicking the convergence of computing and communications that we see in cell phones today. The Internet is sure to evolve, though it remains an open question whether it will retain much of the free-wheeling wild-west character that was so productive in its early years. It might become more constrained and controlled by business and government, a set of "walled gardens"—attractive, to be sure, but walled nevertheless. My bet is the latter, unfortunately; we've already seen examples where entire countries routinely restrict Internet access or cut it off entirely in times of unrest.

Finally, the *universal availability of digital systems*. Digital devices will continue to get smaller, cheaper, faster and more pervasive as they incorporate technological improvements. Improvements in a single technology like storage density often have an effect on all digital devices. The Internet of Things will be all around us as more and more of our devices contain computers and are networked; this will make security problems worse.

The core limitations and likely problems of digital technology will remain operative, and you should be aware of them. Technology contributes many positive things, but it raises new forms of difficult issues and exacerbates existing ones. Here are some of the most important.

Misinformation, disinformation, and fake news of all sorts are a rapidly growing concern on the Internet. False and misleading news stories, images, videos, and the like are rampant on social media sites, which have been entirely too passive about reining in dangerously wrong content. There is certainly a valid concern about censorship and interference with free speech, but in my opinion the pendulum is too far to one side. As one random example, in one 3-month period during the Covid-19 pandemic in 2020, Facebook removed seven million posts that offered "fake preventative measures or exaggerated cures that the CDC and other health experts tell us are dangerous." It also put warnings on nearly 100 million other posts.

Privacy is under continuous threat from attempts to subvert it for commercial, governmental and criminal purposes. Extensive collection of our personal data will continue apace; personal privacy is likely to diminish even further than it already has. The Internet originally made it too easy to be anonymous, especially for bad practices, but today it is almost impossible to remain anonymous even with good intentions. Attempts by governments to control their citizens' access to the Internet and to weaken cryptography will not help the good guys, but they will provide the bad guys with aid, comfort, and a single point of failure to exploit. One might cynically say that governments want their own citizens to be easy to identify and monitor, but support the privacy and anonymity of dissidents in other countries. Businesses are eager to know as much as possible about current and potential customers. Once information is on the web, it's there forever; there's no practical way to call it back.

Surveillance, from ubiquitous cameras to web tracking to recording where our phones are, continues to increase, and the exponentially decreasing cost of storage and processing makes it more and more feasible to keep complete digital records of our entire lives. How much disk space would it take to record everything you have heard and said so far in your life, and how much would that storage cost? If you're 20 years old, the answer is about 10 TB, which in 2021 would cost less than $200. A complete video record would not be more than a factor of 10 or 20 larger.

Security for individuals, corporations and governments is also an ongoing problem. I'm not sure that terms like cyber-warfare, or indeed cyber-anything, are helpful, but it is certain that individuals and larger groups are potentially and often actually under some kind of cyber-attack by nation states and organized criminals. Poor security practices make us all vulnerable to theft of information from government and commercial databases.

Copyright is difficult in a world where it is possible to make unlimited copies of digital material and distribute them throughout the world at zero cost. Traditional copyright functioned acceptably well for creative works before the digital era, because manufacturing and distribution of books, music, movies and TV shows required expertise and specialized equipment. Those days are gone. Copyright and fair use are being replaced by licensing and digital rights management, which don't impede true pirates, though they do inconvenience ordinary people. How do we prevent manufacturers from using copyright to reduce competition and create customer lock-in? How do we protect the rights of authors, composers, performers, film makers, and programmers, while ensuring that their works are not restricted forever?

Patents are also a difficult issue. As more and more devices contain general-purpose computers controlled by software, how do we protect the legitimate interests of innovators while preventing extortion by the holders of too broad or inadequately researched patents?

Resource allocation, particularly of scarce but valuable resources like spectrum, is always going to be contentious. Incumbents—those who already have the allocation, like big telecom companies—have a great advantage here, and they can use their position to maintain it through money, lobbying and natural network effects.

Antitrust is a significant issue in the EU and the US. Companies like Amazon, Facebook and Google dominate their markets, and this gives them outsize concentrated power. Google is perhaps the most vulnerable to antitrust actions; the US Department of Justice announced an antitrust suit against Google late in 2020. At least 70 percent of worldwide searches are made through Google (90 percent in the US). It is the most important company in advertising, which provides most of its income. The large majority of phones run Google's Android operating system. Facebook dominates social media, both directly and through subsidiaries like Instagram. Both Facebook and Google regularly buy small companies to acquire technology and expertise, but also to eliminate potential competition before it grows. Large tech companies argue that they are successful because they provide better services than their competitors do, and their success is a result. But it's also possible to argue that they have too much power, whether legitimately or not. It appears that both the EU and the US are starting to worry about this, and in some cases even take action to control the power of such companies.

Jurisdiction is also difficult in a world where information can travel everywhere. Business and social practices that are legal in one jurisdiction may be illegal in others. Legal systems have not caught up with this at all. The problem is seen in issues like taxes across state borders in the US, and in conflicting data privacy rules in the EU and the US. It's also seen in forum shopping, where plaintiffs start legal actions like patent or libel suits in jurisdictions where they expect favorable outcomes, regardless of where the offense may have occurred or where the defendants might be. Internet jurisdiction itself is under threat from entities that want more control for their own interests.

Control is perhaps the largest issue of all. Governments want to control what their citizens can say and do on the Internet, which of course is increasingly a synonym for all media; country firewalls are likely to become more common and harder to evade. Countries will impose more and more restrictions on what companies must do to stay in business within the country. Companies want their customers confined to walled gardens that are difficult to escape. Think of how many of the devices you use are locked down by their suppliers so you can't run your own software on them or even be sure of what they do. Individuals would like to limit the reach of both governments and corporations, but the playing fields are far from level. The defenses discussed above are a help but in no way sufficient.

Finally, one must always remember that although technology is changing rapidly, people are not. In most respects we're the same as we were thousands of years ago, with similar proportions of good and bad people operating from good and bad motives. Social, legal and political mechanisms do adapt to technological changes, but it can be a slow process and it moves at different speeds and comes to different solutions in different parts of the world. I don't know how things will evolve over the next few years, but I hope that this book will help you to anticipate, cope with, and positively influence some of the inevitable changes.

Notes

"Your organization is sound, your selection of material judicious, and your writing is good. You have not quite grasped the essentials of footnoting. C+."

Grader's comments on an essay I wrote as a junior in university, 1963.

This section collects notes on sources (though by no means complete), including books that I have enjoyed and think you might also like. As always, Wikipedia is an excellent source for a quick survey and basic facts for almost any topic. Search engines do a fine job of locating related material. I have not tried to provide direct links for information that is readily found online. Links were correct at time of publication but may have suffered link-rot.

xi: The IBM 7094 had about 150 KB of RAM, a clock speed of 500 KHz, and cost nearly $3 million: en.wikipedia.org/wiki/IBM_7090.

xiii: Richard Muller, *Physics for Future Presidents*, Norton, 2008. An excellent book, and one of the inspirations for this one.

xiii: Hal Abelson, Ken Ledeen, Harry Lewis, Wendy Seltzer, *Blown to Bits: Your Life, Liberty, and Happiness After the Digital Explosion*, Second edition, Addison-Wesley, 2020. Touches on many important social and political topics, especially about the Internet. Bits and pieces, so to speak, would make good material for my Princeton course, and it derives from an analogous course at Harvard.

2: Zoom's stock price took a significant hit when the FTC accused it of lying about end-to-end encryption, though it subsequently recovered most of the loss.

3: China's Covid app: www.nytimes.com/2020/03/01/business/china-coronavirus-surveillance.html

3: Bruce Schneier's take on the inefficacy of contract-tracing apps: www.schneier.com/blog/archives/2020/05/me_on_covad-19_.html.

3: Snowden's story is told in Glenn Greenwald's *No Place to Hide* (2014), Laura Poitras's prize-winning documentary *Citizenfour* (2015), Snowden's own *Permanent Record* (2019), and Bart Gellman's *Dark Mirror* (2020).

3: www.npr.org/sections/thetwo-way/2014/03/18/291165247/report-nsa-can-record-store-phone-conversations-of-whole-countries.

4: James Gleick, *The Information: A History, A Theory, A Flood*, Pantheon, 2011. Interesting material on communications systems, focusing on Claude Shannon, the father of information theory. The historical parts are especially intriguing.

5: NSA advice on limiting location data: media.defense.gov/2020/Aug/04/2002469874/-1/-1/0/CSI_LIMITING_LOCATION_DATA_EXPOSURE_FINAL.PDF

6: Bruce Schneier, *Data and Goliath: The Hidden Battles to Collect Your Data and Control Your World*, Norton, 2015 (p. 127). Authoritative, disturbing, well written. It's likely to make you justifiably angry.

7: James Essinger, *Jacquard's Web: How a Hand-loom Led to the Birth of the Information Age*, Oxford University Press, 2004. Follows Jacquard's loom through Babbage, Hollerith and Aiken.

7: The Difference Engine picture is a public domain image from Wikipedia: commons.wikimedia.org/wiki/File:Babbage_Difference_Engine_(1).jpg.

8: Doron Swade, *The Difference Engine: Charles Babbage and the Quest to Build the First Computer*, Penguin, 2002. Swade also describes the construction of one of Babbage's machines in 1991, now housed in London's Science Museum; a 2008 clone (Figure I.1, page 8) is in the Computer History Museum in Mountain View, California. See also www.computerhistory.org/babbage.

8: The quotation about music composition is from Ada Lovelace's translation and notes on Luigi Menabrea's "Sketch of the Analytical Engine," 1843.

9: Stephen Wolfram, creator of Mathematica, wrote a long and informative blog post on Lovelace's history: writings.stephenwolfram.com/2015/12/untangling-the-tale-of-ada-lovelace.

9: The Ada Lovelace portrait is a public domain image from Wikipedia: commons.wikimedia.org/wiki/File:Carpenter_portrait_of_Ada_Lovelace_-_detail.png.

9: Scott McCartney, *ENIAC: The Triumphs and Tragedies of the World's First Computer*, Walker & Company, 1999.

11: Burks, Goldstine and von Neumann, "Preliminary discussion of the logical design of an electronic computing instrument," www.cs.unc.edu/~adyilie/comp265/vonNeumann.html.

11: macOS is the current name for Apple's operating system, previously known as Mac OS X.

17: Online copy of *Pride and Prejudice*: www.gutenberg.org/ebooks/1342.

20: Charles Petzold, *Code: The Hidden Language of Computer Hardware and Software*, Microsoft Press, 2000. How computers are built from logic gates; it covers a level or two below what this book does.

22: Gordon Moore, "Cramming more components onto integrated circuits," newsroom.intel.com/wp-content/uploads/sites/11/2018/05/moores-law-electronics.pdf.

27: Excellent explanation of how digital cameras work: www.irregularwebcomic.net/3359.html

37: Leibnitz explored binary and even hexadecimal in the 1670s; he used musical notes (ut, re, mi, fa, sol, la) for the six extra digits.

37: colornames.org is a fun site that illustrates just how many colors 16 million is.

38: In 2020, Apple's Catalina version of macOS no longer supports 32-bit programs.

38: Donald Knuth, *The Art of Computer Programming*, *Vol 2: Seminumerical Algorithms*, Section 4.1, Addison-Wesley, 1997.

50: A Turing machine is an abstract model of computation; there's a marvelous concrete implementation at www.youtube.com/watch?v=E3keLeMwfHY.

51: Alan Turing, "Computing machinery and intelligence." *The Atlantic* has an informative and entertaining article on the Turing test at www.theatlantic.com/magazine/archive/2011/03/mind-vs-machine/8386.

51: The CAPTCHA is a public domain image from en.wikipedia.org/wiki/File:Modern-captcha.jpg.

51: Turing home page maintained by Andrew Hodges: www.turing.org.uk/turing. Hodges is the author of the definitive biography *Alan Turing: The Enigma*. Updated edition, Princeton University Press, 2014.

51: ACM Turing Award: amturing.acm.org/.

51: A 1944 Enigma used by the German navy was sold at auction for $437,000 in 2020: www.zdnet.com/article/rare-and-hardest-to-crack-enigma-code-machine-sells-for-437000.

53: One of many articles on the end of Moore's Law: https://www.technologyreview.com/2020/02/24/905789/were-not-prepared-for-the-end-of-moores-law/.

56: A description of the 737 MAX situation from a software perspective: spectrum.ieee.org/aerospace/aviation/how-the-boeing-737-max-disaster-looks-to-a-software-developer.

56: The Iowa Democratic primary fiasco: www.nytimes.com/2020/02/09/us/politics/iowa-democratic-caucuses.html.

56: The perils of Internet voting, triggered by coronavirus concerns: www.politico.com/news/2020/06/08/online-voting-304013

56: www.cnn.com/2016/02/03/politics/cyberattack-ukraine-power-grid.

57: en.wikipedia.org/wiki/WannaCry_ransomware_attack.

57: thehill.com/policy/national-security/507744-russian-hackers-return-to-spotlight-with-vaccine-research-attack

59: James Gleick on Richard Feynman: "Part Showman, All Genius," www.nytimes.com/1992/09/20/magazine/part-showman-all-genius.html.

59: The River Cafe Cookbook, "The best chocolate cake ever," books.google.com/books?id=INFnzXj81-QC&pg=PT512.

68: William Cook's *In Pursuit of the Traveling Salesman*, Princeton University Press, 2011, is an engaging description of the history and the state of the art.

69: A 2013 episode of the series *Elementary* centers on P=NP: www.imdb.com/title/tt3125780/.

70: John MacCormick's *Nine Algorithms That Changed the Future: The Ingenious Ideas That Drive Today's Computers*, Princeton University Press, 2011, provides an accessible description of some major algorithms, including search, compression, error correction and cryptography.

77: Kurt Beyer, *Grace Hopper and the Invention of the Information Age*, MIT Press, 2009. Hopper was a remarkable figure, a computing pioneer of great influence, and at her retirement at 79, the oldest commissioned officer in the US Navy. One of her set pieces in speeches was to hold out her hands a foot apart and say "That's a nanosecond."

82: NASA Mars Climate Orbiter report: llis.nasa.gov/llis_lib/pdf/1009464main1_0641-mr.pdf.

82: www.wired.com/2015/09/google-2-billion-lines-codeand-one-place.

84: The bug picture is a public domain image from www.history.navy.mil/our-collections/photography/numerical-list-of-images/nhhc-series/nh-series/NH-96000/NH-96566-KN.html.

85: www.theregister.co.uk/2015/09/04/nsa_explains_handling_zerodays.

87: Supreme Court decision confirming the constitutionality of the 1998 Sonny Bono Copyright Term Extension Act, sarcastically known as the Mickey Mouse Protection Act, because it extended the already long copyright protection of Mickey Mouse and other Disney characters. en.wikipedia.org/wiki/Eldred_v._Ashcroft.

88: Amazon 1-click patent: www.google.com/patents?id=O2YXAAAAEBAJ.

88: Wikipedia has a good discussion of patent trolls: en.wikipedia.org/wiki/Patent_troll.

89: The EULA comes from About this Mac / Support / Important Information... / Software License Agreement. It's about 12 pages long.

89: From the macOS Mojave EULA: "You also agree that you will not use the Apple Software for any purposes prohibited by United States law, including, without limitation, the development, design, manufacture or production of missiles, nuclear, chemical or biological weapons."

91: en.wikipedia.org/wiki/Oracle_America,_Inc._v._Google,_Inc.

93: Code for my car: www.fujitsu-ten.com/support/source/oem/14f.

96: *Unix: A History and a Memoir* (Kindle Direct Publishing, 2019) is my personal take on Unix history, from the perspective of someone present at the creation, though not responsible for it.

99: The original Linux code can be found at www.kernel.org/pub/linux/kernel/Historic.

108: Windows file recovery tool: www.microsoft.com/en-us/p/windows-file-recovery/9n26s50ln705.

109: One example from 65 million returned by a Google search: "Leaked White House emails reveal behind-the-scenes battle over chloroquine in coronavirus response".

113: Microsoft Windows on ARM processors: docs.microsoft.com/en-us/windows/uwp/porting/apps-on-arm.

114: Court's Findings of Fact, paragraph 154, 1999, at www.justice.gov/atr/cases/f3800/msjudgex.htm. The case finally ended in 2011 when the last oversight of Microsoft's compliance ended.

115: Obama's exhortation, on YouTube, was part of a Computer Science Education Week campaign: www.whitehouse.gov/blog/2013/12/09/don-t-just-play-your-phone-program-it.

124: jsfiddle.net and w3schools.com are two of many useful sites for learning JavaScript.

125: You can download Python from python.org.

125: Colab is accessible at colab.research.google.com.

125: A Jupyter notebook is "an open-source web application that allows you to create and share documents that contain live code, equations, visualizations and narrative text." See jupyter.org.

136: Gerard Holzmann and Bjorn Pehrson, *The Early History of Data Networks*, IEEE Press, 1994. Detailed and interesting history of the optical telegraph.

136: The optical telegraph drawing is a public domain image from en.wikipedia.org/wiki/File:Telegraph_Chappe_1.jpg.

137: Tom Standage, *The Victorian Internet: The Remarkable Story of the Telegraph and the Nineteenth Century's On-Line Pioneers*, Walker, 1998. Fascinating and entertaining reading.

137: I'm not the only one who misses life before cell phones: www.theatlantic.com/technology/archive/2015/08/why-people-hate-making-phone-calls/401114.

141: The papers of Alexander Graham Bell are online; the quotation comes from memory.loc.gov/mss/magbell/253/25300201/0022.jpg.

142: www.10stripe.com/articles/why-is-56k-the-fastest-dialup-modem-speed.php.

143: A good description of DSL: broadbandnow.com/DSL.

148: Guy Klemens, *Cellphone: The History and Technology of the Gadget that Changed the World*, McFarland, 2010. Detailed history and technical facts on the evolution of cell phones. Some parts are heavy going, but much is accessible; it gives a good picture of the remarkable complexity of a system we take for granted.

149: A US federal judge suppressed evidence from a stingray: www.reuters.com/article/us-usa-crime-stingray-idUSKCN0ZS2VI.

151: For a good explanation of 4G and LTE, see www.digitaltrends.com/mobile/4g-vs-lte.

153: Interactive explanation of how JPEG works: parametric.press/issue-01/unraveling-the-jpeg/

159: NSA and GCHQ are both tapping fiber optic cables where they make landfall: www.theatlantic.com/international/archive/2013/07/the-creepy-long-standing-practice-of-undersea-cable-tapping/277855.

161: RFC on Avian carriers: tools.ietf.org/html/rfc1149. You might also enjoy RFC-2324.

162: The current list of top-level domains is at www.iana.org/domains/root/db; there are nearly 1,600.

162: Law enforcement often fails to realize that an IP address does not definitively identify an individual: www.eff.org/files/2016/09/22/2016.09.20_final_formatted_ip_address_white_paper.pdf.

165: DE-CIX, like many IXPs, provides extensive traffic graphs; see www.de-cix.net.

166: `traceroute` was created by Van Jacobson in 1987.

172: SMTP was originally defined by Jon Postel in RFC 788 in 1981.

172: SMTP session at technet.microsoft.com/en-us/library/bb123686.aspx.

176: In 2015, Keurig tried to enforce DRM on pods for its coffee makers; users were not happy, and sales went down dramatically: boingboing.net/2015/05/08/keurig-ceo-blames-disastrous-f.html

176: Devices call home: www.digitaltrends.com/news/china-spying-iot-devices.

177: arstechnica.com/security/2016/01/how-to-search-the-internet-of-things-for-photos-of-sleeping-babies.

177: Gordon Chu, Noah Apthorpe, Nick Feamster, "Security and Privacy Analyses of Internet of Things Children's Toys," 2019.

177: Using Telnet to access IoT devices: www.schneier.com/blog/archives/2020/07/half_a_million.html.

177: Attacks on wind turbines: news.softpedia.com/news/script-kiddies-can-now-launch-xss-attacks-against-iot-wind-turbines-497331.shtml.

184: Accessibility for visually impaired: www.afb.org/about-afb/what-we-do/afb-consulting/afb-accessibility-resources/improving-your-web-site.

188: Microsoft's *10 Immutable Laws of Security*: docs.microsoft.com/en-us/archive/blogs/rhalbheer/ten-immutable-laws-of-security-version-2-0.

190: Kim Zetter, *Countdown to Zero Day*, Crown, 2014, is a gripping description of Stuxnet.

194: blog.twitter.com/en_us/topics/company/2020/an-update-on-our-security-incident.html.

194: CEO phishing gave away W-2s on all employees at Seagate in 2016: krebsonsecurity.com/2016/03/seagate-phish-exposes-all-employee-w-2s.

195: www.ucsf.edu/news/2020/06/417911/update-it-security-incident-ucsf

196: epic.org/privacy/data-breach/equifax/

197: Wawa's statement on their security breach: www.wawa.com/alerts/data-security.

197: Data breach at Clearview AI: www.cnn.com/2020/02/26/tech/clearview-ai-hack/index.html.

197: news.marriott.com/news/2020/03/31/marriott-international-notifies-guests-of-property-system-incident

197: Amazon DDoS attack: www.theverge.com/2020/6/18/21295337/amazon-aws-biggest-ddos-attack-ever-2-3-tbps-shield-github-netscout-arbor

198: Breach of free "no-logging" VPNs: www.theregister.com/2020/07/17/ufo_vpn_database/

198: The FTC complaint and proposed settlement with Zoom: www.ftc.gov/news-events/press-releases/2020/11/ftc-requires-zoom-enhance-its-security-practices-part-settlement.

198: Steve Bellovin's *Thinking Security*, Addison-Wesley, 2015, has an extensive discussion of threat models.

199: There's a famous xkcd comic on choosing passwords: xkcd.com/936.

200: help.getadblock.com/support/solutions/articles/6000087914-how-does-adblock-work-

201: www.theguardian.com/technology/2020/jan/21/amazon-boss-jeff-bezoss-phone-hacked-by-saudi-crown-prince

201: *Click Here to Kill Everybody*, Bruce Schneier, Norton, 2018.

201: Eli Pariser, *The Filter Bubble: What the Internet Is Hiding from You*, Penguin, 2011.

203: Dr. Seuss's 1955 children's book *On Beyond Zebra!* describes a fanciful extended alphabet.

205: Cisco's prediction is one of several that anticipate greatly increased Internet traffic: www.cisco.com/c/en/us/solutions/collateral/executive-perspectives/annual-internet-report/white-paper-c11-741490.html.

206: The original Google paper: infolab.stanford.edu/~backrub/google.html. The system really was called "BackRub" in its first incarnation.

206: Two sites with big numbers: www.domo.com/learn/data-never-sleeps-5, www.forbes.com/sites/bernardmarr/2018/05/21/how-much-data-do-we-create-every-day-the-mind-blowing-stats-everyone-should-read.

209: Latanya Sweeney discovered that searches for names "generated ads suggestive of an arrest" significantly more often for names that are "racially associated." See papers.ssrn.com/sol3/papers.cfm?abstract_id=2208240.

209: www.reuters.com/article/us-facebook-advertisers/hud-charges-facebook-with-housing-discrimination-in-targeted-ads-on-its-platform-idUSKCN1R91E8.

209: www.propublica.org/article/facebook-ads-can-still-discriminate-against-women-and-older-workers-despite-a-civil-rights-settlement

209: DuckDuckGo's privacy advice can be found at spreadprivacy.com.

210: www.nytimes.com/series/new-york-times-privacy-project

210: www.nytimes.com/interactive/2019/12/19/opinion/location-tracking-cell-phone.html

212: www.washingtonpost.com/news/the-intersect/wp/2016/08/19/98-personal-data-points-that-facebook-uses-to-target-ads-to-you/

212: 98 things that Facebook uses to target you: www.washingtonpost.com/technology/2020/01/28/off-facebook-activity-page.

213: Netflix privacy policy: help.netflix.com/legal/privacy, June 2020.

214: Canvas fingerprinting: en.wikipedia.org/wiki/Canvas_fingerprinting.

215: How to turn off speech-enabled "smart" TVs: www.consumerreports.org/privacy/how-to-turn-off-smart-tv-snooping-features/.

215: www.nytimes.com/2020/07/16/business/eu-data-transfer-pact-rejected.html

216: www.pewresearch.org/internet/2019/01/16/facebook-algorithms-and-personal-data.

216: In 2019, the *New York Times* analyzed 150 privacy policies. "They were an incomprehensible disaster." www.nytimes.com/interactive/2019/06/12/opinion/facebook-google-privacy-policies.html

216: www.swirl.com/products/beacons.

217: Locational privacy: www.eff.org/wp/locational-privacy. The Electronic Frontier Foundation at eff.org is a good source for privacy and security policy information.

217: fas.org/irp/congress/2013_hr/100213felten.pdf.

218: Kosinski et al., "Private traits and attributes are predictable from digital records of human behavior," www.pnas.org/content/early/2013/03/06/1218772110.full.pdf+html.

218: Facebook tagging help: www.facebook.com/help/187272841323203 (June 2020)

221: Simson L. Garfinkel, De-Identification of Personal Information, dx.doi.org/10.6028/NIST.IR.8053

221: georgetownlawtechreview.org/re-identification-of-anonymized-data/GLTR-04-2017.

221: Cloud image courtesy of clipartion.com/free-clipart-549.

223: In April 2016, Microsoft sued the Department of Justice over this kind of requirement: blogs.microsoft.com/on-the-issues/2016/04/14/keeping-secrecy-exception-not-rule-issue-consumers-businesses.

224: www.theguardian.com/commentisfree/2014/may/20/why-did-lavabit-shut-down-snowden-email.

224: A government redaction error revealed that Snowden was the target: www.wired.com/2016/03/government-error-just-revealed-snowden-target-lavabit-case.

225: Transparency reports: www.google.com/transparencyreport, govtrequests.facebook.com, aws.amazon.com/compliance/amazon-information-requests.

228: On the relationship of ML to statistics: www.svds.com/machine-learning-vs-statistics

228: "vas3k.com/blog/machine_learning" by Vasily Zubarev is an excellent informal introduction with good illustrations and no mathematics.

229: Computer History Museum retrospective on expert systems (2018): www.computerhistory.org/collections/catalog/102781121.

232: The Turing Award page on Bengio, Hinton and LeCun is at awards.acm.org/about/2018-turing

233: www.nytimes.com/2020/06/24/technology/facial-recognition-arrest.html

233: IBM abandons facial recognition: www.ibm.com/blogs/policy/facial-recognition-susset-racial-justice-reforms.

233: Amazon suspends facial recognition: yro.slashdot.org/story/20/06/10/2336230/amazon-pauses-police-use-of-facial-recognition-tech-for-a-year.

235: The Eliza dialog comes from www.masswerk.at/elizabot.

235: Talk to Transformer can be accessed at inferkit.com.

238: Amazon Rekognition: www.nytimes.com/2020/06/10/technology/amazon-facial-recognition-backlash.html.

238: Clearview AI suit: /www.nytimes.com/2020/08/11/technology/clearview-floyd-abrams.html.

238: Significant ideas for this section come from *Fairness and Machine Learning: Limitations and Opportunities*, a book by Barocas, Hardt and Narayanan (fairmlbook.org).

238: Botpoet.com is an entertaining online Turing test for poetry.

240: Simon Singh, *The Code Book*, Anchor, 2000. A delightful history of cryptography for the general reader. The Babington Plot (the attempt to put Mary, Queen of Scots, on the throne) is fascinating.

241: The Enigma machine photograph is a public domain image from Wikipedia: commons.wikimedia.org/wiki/File:EnigmaMachine.jpg.

242: Bruce Schneier has several essays on why amateur cryptography doesn't work; this one points to earlier ones as well: www.schneier.com/blog/archives/2015/05/amateurs_produc.html.

242: Ronald Rivest says that "It seems highly likely that this standard was designed by the NSA to explicitly leak users' key information to the NSA (and to no one else). The Dual-EC-DRBG standard apparently (and I would suggest, almost certainly) contains a "back-door" enabling the NSA to have surreptitious access." www.nist.gov/public_affairs/releases/upload/VCAT-Report-on-NIST-Cryptographic-Standards-and-Guidelines-Process.pdf.

243: Alice, Bob and Eve at xkcd.com/177.

245: When signing and encryption are combined, the inner crypto layer must somehow depend on the outer layer, so as to reveal any tampering with the outer layer. world.std.com/˜dtd/sign_encrypt/sign_encrypt7.html

247: Snapchat privacy policy: www.snapchat.com/privacy.

249: A list of things *not* to do when using Tor: www.whonix.org/wiki/DoNot.

249: www.washingtonpost.com/news/the-switch/wp/2013/10/04/everything-you-need-to-know-about-the-nsa-and-tor-in-one-faq.

249: The Snowden documents can be found at www.aclu.org/nsa-documents-search and www.cjfe.org/snowden, among others.

249: TAILS web site: tails.boum.org.

250: Bitcoin historical prices are from Yahoo Finance.

251: Some people whose identities were found on the hacked Ashley Madison site for extramarital affairs received blackmail demands for $2000 in bitcoins: www.grahamcluley.com/2016/01/ashley-madison-blackmail-letter.

252: www.irs.gov/individuals/international-taxpayers/frequently-asked-questions-on-virtual-currency-transactions

252: Arvind Narayanan et al., *Bitcoin and Cryptocurrency Technologies*, Princeton University Press, 2016.

252: "Keys under doormats": dspace.mit.edu/handle/1721.1/97690. The authors are a truly knowledge-able group of cryptography experts. I know half of these people personally, and trust their expertise and motives.

253: www.nytimes.com/2020/01/07/technology/apple-fbi-iphone-encryption.html

257: www.msn.com/en-us/news/technology/facebook-says-it-removed-over-7m-pieces-of-wrong-covid-19-content-in-quarter/ar-BB17Q4qu.

Glossary

"Some words there are which I cannot explain, because I do not understand them."

Samuel Johnson, *A Dictionary of the English Language*, 1755.

The glossary provides brief definitions or explanations of important terms that appear in the book, focusing on ones that use ordinary words but with special meanings, and that you are likely to see frequently.

Computers and communications systems like the Internet deal with very large numbers, often expressed in terms of unfamiliar units. The table below defines all the units that appear in the book, along with the others of the International System of Units. As technology advances, you'll see more of the ones that represent big numbers. The table also shows the nearest powers of two. The error is only 21 percent at 10^{24}; that is, 2^{80} is about 1.21×10^{24}.

SI name	power of 10	common name	nearest power of 2
yocto	10^{-24}		2^{-80}
zepto	10^{-21}		2^{-70}
atto	10^{-18}		2^{-60}
femto	10^{-15}		2^{-50}
pico	10^{-12}	trillionth	2^{-40}
nano	10^{-9}	billionth	2^{-30}
micro	10^{-6}	millionth	2^{-20}
milli	10^{-3}	thousandth	2^{-10}
-	10^{0}		2^{0}
kilo	10^{3}	thousand	2^{10}
mega	10^{6}	million	2^{20}
giga	10^{9}	billion	2^{30}
tera	10^{12}	trillion	2^{40}
peta	10^{15}	quadrillion	2^{50}
exa	10^{18}	quintillion	2^{60}
zetta	10^{21}		2^{70}
yotta	10^{24}		2^{80}

4G Fourth generation, a somewhat imprecise term characterizing the technology used in smartphones, roughly 2010 onward; the successor to 3G.

5G Fifth generation, newer and more precisely defined, roughly 2020 onward; the replacement for 4G.

802.11 The standard for wireless systems like those used in laptops and home routers; also Wi-Fi.

add-on A small JavaScript program added to a browser for extra features or convenience; privacy add-ons like Adblock Plus and NoScript are examples. Also called extension.

AES Advanced Encryption Standard, the most widely used secret-key encryption algorithm.

algorithm A precise and complete specification of a computational process, but abstract and not directly executable by a computer, in contrast to a program.

AM Amplitude Modulation, a mechanism for adding information like voice or data to a signal by modifying the signal amplitude; usually seen in the context of AM radio. See FM.

analog Generic term for representation of information that uses a physical property that varies smoothly in proportion, such as the level of liquid in a thermometer; contrast with digital.

API Application Programming Interface, a description for programmers of services provided by a library or other collection of software; for instance the Google Maps API describes how to control map displays with JavaScript.

app, application A program or family of programs that perform some task, for example Word or iPhoto; *app* is most often used for cell phone applications like calendars and games. "Killer app" is an earlier use.

architecture An imprecise word for the organization or structure of a computer program, system or hardware.

ASCII American Standard Code for Information Interchange, a 7-bit encoding of letters, digits, and punctuation; almost always stored as 8-bit bytes.

assembler A program that translates instructions in the processor's repertoire into bits for loading directly into the memory of a computer; *assembly language* is the corresponding level of programming language.

backdoor In cryptography, an intentional weakness that permits someone with additional knowledge to break or circumvent encryption.

bandwidth How fast a communications path can carry information, measured in bits per second (bps), for instance 56 Kbps for a telephone modem or 100 Mbps for Ethernet.

base station Radio equipment that connects wireless devices (cell phones, laptops) to a network (telephone network, computer network).

binary Having only two states or possible values; also *binary number* for numbers in base 2.

binary search An algorithm that searches a sorted list, by repeatedly dividing the part to be searched next into two equal halves.

bit A binary digit (0 or 1) that represents the information in a binary choice like on or off.

Bitcoin A digital or cryptocurrency that allows anonymous online transactions using peer-to-peer networking.

BitTorrent Peer-to-peer protocol for distributing large popular files efficiently; downloaders also upload.

blockchain The distributed ledger of all previous transactions used by the Bitcoin protocol.

Bluetooth Short-range low-power radio for hands-free phones, games, keyboards, and the like.

bot, botnet A computer running a program under the control of a bad guy; a botnet is a collection of bots under a common control. From robot.

browser A program like Chrome, Firefox, Internet Explorer, Edge or Safari that provides the primary interface to web services for most people.

browser fingerprinting A technique by which a server can use properties of a user's browser to identify that user more or less uniquely. *Canvas fingerprinting* is one mechanism.

bug An error in a program or other system.

bus A set of wires used to connect electronic devices; also see USB.

byte Eight bits, enough to store a letter, a small number, or part of a larger quantity; treated as a unit in modern computers.

cable modem A device for sending and receiving digital data on a cable television network.

cache   Local storage that provides fast access to information that has been used recently.

CAPTCHA   Test to distinguish humans from computers; intended to detect bots.

certificate   Digitally signed cryptographic data that can be used to verify the authenticity of a web site.

chip   Small electronic circuit, manufactured on a flat silicon surface and mounted in a ceramic package; also integrated circuit, microchip.

Chrome OS   An operating system from Google on which applications and user data primarily live in the cloud rather than on a local machine, and are accessed by a browser.

client   A program, often a browser, that makes requests of a server, as in client-server.

cloud computing   Computing performed on a server, with data stored on a server, replacing a desktop application; mail, calendar and photo sharing sites are examples.

code   Text of a program in a programming language, as in source code; an encoding, as in ASCII.

compiler   A program that translates programs written in a high-level language like C or Fortran into a lower-level form like assembly language.

complexity   A measure of the difficulty of a computational task or algorithm, expressed in terms of how long it takes to process N data items, like N or $\log N$.

compression   Squeezing a digital representation into fewer bits, as in MP3 compression of digital music or JPEG compression of images.

cookie   Text sent by a server, stored by the browser on your computer, and then returned by your browser on your next access to that server; widely used for tracking visits to web sites.

CPU   Central Processing Unit; see processor.

cryptocurrency   Digital currency (like Bitcoin) based on cryptographic techniques, not physical assets or government fiat.

dark web   Part of the World Wide Web that is accessible only with special software and/or access information; largely associated with illegal activities.

declaration   A programming language construct that states the name and properties of some part of a computer program, for example a variable that will store information during a computation.

deep learning   Machine learning technique based on networks of artificial neurons.

deprecated   In computing, indicates a technology that is going to be replaced or made obsolete and thus should be avoided.

DES   Data Encryption Standard, the first widely used digital encryption algorithm; superseded by AES.

digital   Representation of information that takes on only discrete numeric values; contrast to analog.

directory   Same as folder.

DMCA   Digital Millennium Copyright Act (1998), the US law that protects copyrighted digital material.

DNS   Domain Name System, the Internet service that translates domain names into IP addresses.

domain name   A hierarchical naming scheme for computers connected to the Internet, such as www.cs.nott.ac.uk.

driver   Software that controls a particular hardware device like a printer; usually loaded into the operating system as necessary.

DRM   Digital Rights Management, techniques for preventing illegal copying of copyrighted material; generally unsuccessful.

DSL   Digital Subscriber Loop, a technique of sending digital data over telephone lines. Comparable to cable, but less often used.

Ethernet   The most common local area network technology, used in most home and office wireless networks.

EULA   End User License Agreement, the long legal document in tiny print that restricts what you can do with software and other digital information.

exponential   Growing by a fixed proportion each fixed step size or time period, for example, growing by 6 percent a month; often used carelessly for "growing quickly."

fiber, optical fiber   Fine strand of extremely pure glass used to carry light signals over long distances; the signals encode digital information. Most long-distance digital traffic is carried on fiber optic cables.

file system   The part of an operating system that organizes and accesses information on disks and other secondary storage media.

filter bubble   The narrowing of sources and information that results from relying on restricted sources of online information.

firewall   A program and perhaps hardware that controls or blocks incoming and outgoing network connections from a computer or a network.

Flash   Adobe software system for displaying video and animation on web pages; deprecated.

flash memory   Integrated-circuit memory technology that preserves data without consuming power; used in cameras, phones, USB memory sticks and as a replacement for disk drives.

FM   Frequency Modulation, a technique for sending information by changing the frequency of a radio signal; usually seen in the context of FM radio. See AM.

folder   A file that holds information about files and folders, including size, date, permissions, and location; same as directory.

function   Component of a program that performs a specific focused computational task, for instance computing a square root or popping up a dialog box, like the `prompt` function in JavaScript.

gateway   A computer that connects one network to another; often called a router.

GDPR   General Data Protection Regulation, a European Union law to give individuals control over their online data.

GIF   Graphics Interchange Format, a compression algorithm for simple images with blocks of color, but not photographs. See JPEG, PNG.

GNU GPL   GNU General Public License, a copyright license that protects open-source code by requiring free access to source code, thus preventing it from being taken private.

GPS   Global Positioning System; uses time signals from satellites to compute position on the surface of the earth. It's one-way; GPS devices like car navigators do not broadcast to the satellites.

GSM   Global System for Mobile Communications, a cell phone system used in much of the world.

hard disk   Device that stores data on rotating disks of magnetic material; also hard drive. Contrast with floppy disk.

hexadecimal   Base 16 notation, most often seen in Unicode tables, URLs and color specifications.

HTML   Hypertext Markup Language; used to describe the contents and format of a web page.

HTTP, HTTPS   Hypertext Transfer Protocol; used between clients like browsers and servers; HTTPS is encrypted end to end and thus comparatively secure against snooping and man-in-the-middle attacks.

IC, integrated circuit   Electronic circuit component fabricated on a flat surface, mounted in a sealed package, and connected to other devices in a circuit. Most digital devices are made up mostly of IC's.

ICANN   Internet Corporation for Assigned Names and Numbers, the organization that allocates Internet resources that must be unique, like domain names and protocol numbers.

intellectual property   The product of creative or inventive acts, protectable by copyright and patents; it includes software and digital media. Sometimes confusingly abbreviated IP.

interface   Vague general term for the boundary between two independent entities. See API for programming interfaces. Another use is *(graphical) user interface* or GUI, the part of a computer program that a human interacts with directly.

interpreter   Program that interprets instructions for a computer, whether real or not, thus simulating its behavior; JavaScript programs in a browser are processed by an interpreter. See also virtual machine.

IP   Internet Protocol, the fundamental protocol for sending packets through the Internet; may instead refer to intellectual property.

IP address   Internet Protocol address, the unique numeric address currently associated with a computer on the Internet; loosely analogous to a telephone number.

IPv4, IPv6   The two versions of the IP protocol; IPv4 uses 32-bit addresses, IPv6 uses 128. There are no other versions.

ISP   Internet Service Provider, an entity that provides connections to the Internet; examples include universities, and cable and telephone companies.

IXP   Internet Exchange Point, a physical site where multiple networks meet and data is exchanged between them.

JavaScript   A programming language primarily used on web pages for visual effects and tracking.

JPEG   A standard compression algorithm and representation for digital images, named after the Joint Photographic Experts Group.

kernel The central part of an operating system, responsible for controlling operation and resources.

key logger Software that records all keystrokes on a computer, usually for nefarious purposes.

library A collection of related software components in a form that can be used as parts of a program, for example the standard functions that JavaScript provides for accessing the browser.

Linux An open-source Unix-like operating system, widely used on servers.

logarithm Given a number N, the power to which the base is raised to produce N. In this book, the base is 2 and the logarithms are integers.

loop A part of a program that repeats a sequence of instructions; an infinite loop repeats them a lot of times.

malware Software with malicious properties and intent.

man-in-the-middle attack An attack where an adversary intercepts and modifies communications between two other parties.

microchip Another word for chip or integrated circuit.

modem Modulator / demodulator, a device that converts bits into an analog representation (like sound) and back.

MD5 A message digest or cryptographic hash algorithm; deprecated.

MP3 A compression algorithm and representation for digital audio, part of the MPEG standard for video.

MPEG A standard compression algorithm and representation for digital video, names after the Moving Picture Experts Group.

net neutrality The general principle that Internet service providers should treat all traffic the same way (except perhaps in cases of overload), rather than biasing treatment for economic or other non-technical reasons.

neural network Network of artificial neurons loosely like neurons in the brain, used in machine learning algorithms.

object code Instructions and data in binary form that can be loaded into primary memory for execution; the result of compilation and assembly. Contrast with source code.

open source Source code (that is, readable by programmers) that is freely available, usually under a license like the GNU GPL that keeps it freely available on the same terms.

operating system Program that controls the resources of a computer, including processor, file system, devices, and external connections; examples include Windows, macOS, Unix, Linux.

packet A collection of information in a specified format, such as an IP packet; loosely analogous to a standard envelope or shipping container.

PDF Portable Document Format, a standard representation for printable documents, originally created by Adobe.

peer-to-peer Exchange of information between peers, that is, a symmetric relationship, in contrast to client-server. Used for file-sharing networks and bitcoin.

peripheral Hardware device connected to a computer, like an external disk, printer or scanner.

phishing, spear phishing Attempt, usually by email, to obtain personal information or induce the target to download malware or reveal credentials by pretending to have some kind of relationship with the target; *spear phishing* is more precisely targeted.

pixel Picture element; a single point in a digital image.

platform Vague term for a software system, like an operating system, that provides services that can be built upon.

plug-in A program that runs in the context of a browser; Flash and QuickTime are common examples.

PNG Portable Network Graphics, a lossless compression algorithm, a non-patented replacement for GIF, supporting many more colors; used for text, line art, and images with large areas of solid color.

processor The part of the computer that does arithmetic and logic, and controls the rest of the computer; also CPU. Intel and AMD processors are widely used in laptops; ARM processors are used in most phones.

program A set of instructions that causes a computer to do a task; written in a programming language.

programming language Notation for expressing sequences of operations for a computer, translated ultimately into bits to be loaded into RAM; examples include assembler, C, C++, Java, JavaScript.

protocol An agreement on how systems interact; most often seen in the Internet, which has a large number of protocols for exchanging information over networks.

quadratic Numeric growth proportional to the square of a variable or parameter, for instance how the running time of selection sort varies with the number of items to be sorted, or the area of a circle with the radius.

RAM Random Access Memory; the primary memory in a computer.

ransomware An attack that encrypts data on the victim's computer, requiring a payment to recover it.

registrar A company that has the authority (from ICANN) to sell domain names to individuals and companies.

reinforcement learning Machine learning that uses performance on a real world task to guide learning; used in computer games like chess.

representation General word for how information is expressed in digital form.

RFID Radio-Frequency Identification, a low power wireless system used in electronic door locks, pet identification chips, and the like.

RGB Red, Green, Blue, the standard way that colors are represented in computer displays as the combination of three basic colors.

router Another word for gateway: a computer that passes information from one network to another; also wireless router.

RSA The most widely used public-key encryption algorithm, named after its inventors Ron Rivest, Adi Shamir and Leonard Adleman.

SDK Software Development Kit, a collection of tools to help programmers write programs for some device or environment, for example cell phones and game consoles.

search engine Server like Bing or Google that collects web pages and answers queries about them.

server Computer or computers that provide access to data upon request from a client; search engines, shopping sites, and social networks are examples.

SHA-1, SHA-2, SHA-3 Secure hash algorithms, for making cryptographic digests of arbitrary input; SHA-1 is deprecated.

simulator Program that simulates (acts like) a device or other system.

smartphone Phone like iPhone and Android with the capability of downloading and running programs (apps).

social engineering Technique of deceiving a victim into releasing information or doing some action by pretending to have a relationship like a mutual friend or the same employer.

solid state disk/drive, SSD Non-volatile secondary storage that uses flash memory; a replacement for hard disk drives based on rotating machinery.

source code Program text written in a language comprehensible by programmers, to be compiled into object code.

spectrum Frequency range for a system or device, for example phone service or a radio station.

spyware Software that reports back to home on what happens on the computer where it is installed.

stingray Device that simulates a cell phone base station so phones will communicate with it instead of the regular phone system.

supervised learning Machine learning based on learning from a set of labeled or tagged examples.

system call Mechanism by which an operating system makes its services available to programmers; a system call looks much like a function call.

standard Formal specification or description of how something works or is built or is controlled, precise enough to permit interoperability and independent implementations. Examples include character sets like ASCII and Unicode, plugs and sockets like USB, and programming language definitions.

TCP Transmission Control Protocol, a protocol that uses IP to create two-way streams. TCP/IP is the combination of TCP and IP.

tracking Recording the sites that a web user visits and what he or she does there.

Trojan horse A program that promises to do one thing but in fact does something different and usually malicious.

troll Intentionally disruptive in the Internet; both noun and verb. Also patent troll, who seeks to exploit sketchy patents.

Turing machine   Abstract computer, conceived by Alan Turing, that is capable of performing any digital computation; a universal Turing machine can simulate any other Turing machine and thus any digital computer.

Unicode   A standard encoding for all of the characters in all of the world's writing systems. UTF-8 is an 8-bit variable-width encoding for transferring Unicode data.

Unix   An operating system developed at Bell Labs that forms the base of many of today's operating systems; Linux is a lookalike that provides the same services but with a different implementation.

unsupervised learning   Machine learning based on learning without labeled or tagged examples.

URL   Uniform Resource Locator, the standard form of a web address, like http://www.amazon.com.

USB   Universal Serial Bus, a standard connector for plugging devices like external disk drives, cameras, displays and phones into computers. USB-C is a newer and physically incompatible version.

variable   A RAM location that stores information; a variable declaration names the variable and may provide other information about it, like an initial value or the type of data it holds.

virtual machine   A program that simulates a computer; also interpreter.

virtual memory   Software and hardware that give the illusion of unlimited primary memory.

virus   A program, usually malicious, that infects a computer; a virus needs help propagating from one system to another, in contrast to a worm.

VoIP   Voice over IP, a method of using the Internet for voice conversations, often with a way to access the regular telephone system.

VPN   Virtual Private Network, an encrypted path among computers that secures information flow in both directions.

walled garden   A software ecosystem that confines its users to facilities of that system, making it hard to access or use anything outside the system.

web beacon   A small and usually invisible image used for tracking the fact that a particular web page has been downloaded.

web server   A server focused on web applications.

Wi-Fi   Wireless Fidelity, the marketing name for 802.11 wireless.

wireless router   A radio device that connects wireless devices like computers to a wired network.

worm   A program, usually malicious, that infects a computer; a worm can propagate from one system to another without help, in contrast to a virus.

zero-day   A software vulnerability which defenders have zero days to fix or defend against.

Index

"Any inaccuracies in this index may be explained by the fact that it has been sorted with the help of a computer."

Donald E. Knuth, *The Art of Computer Programming*,
Volume 3, *Searching and Sorting*, 1973.